Science and Technology Policy:
Perspectives for the 1980s

ANNALS OF THE NEW YORK ACADEMY OF SCIENCES

VOLUME 334

Science and Technology Policy: Perspectives for the 1980s

Edited by
HERBERT I. FUSFELD *and* CARMELA S. HAKLISCH

THE NEW YORK ACADEMY OF SCIENCES
NEW YORK, NEW YORK
1979

Library of Congress Cataloging in Publication Data

Main entry under title:

Science and technology policy.

(Annals of the New York Academy of Sciences; v. 334)
Papers presented at a conference held Mar. 1979.
1. Science and state—Congresses. 2. Technology and state—Congresses. I. Fusfeld, Herbert I. II. Haklisch, Carmela S. III. Series: New York Academy of Sciences. Annals; v. 334

Q11.N5 vol. 334 [Q124.6] 500s [351.85'5] 79-28295

CCP
Printed in the United States of America
ISBN 0-89766-036-6

ANNALS OF THE NEW YORK ACADEMY OF SCIENCES

VOLUME 334

December 14, 1979

SCIENCE AND TECHNOLOGY POLICY: PERSPECTIVES FOR THE 1980s*

Conference Chairman
HERBERT I. FUSFELD

Conference Editors
HERBERT I. FUSFELD AND CARMELA S. HAKLISCH

━━━━━━

CONTENTS

* The material in this volume is based upon presentations from a conference entitled International Conference on Science and Technology Policy held March 27 and 28, 1979 by New York University, which was partially supported by the National Science Foundation under Grant No. PRA-7909060. Any opinions, findings, and conclusions or recommendations expressed in this publication are those of the authors and participants and do not necessarily reflect the views of the National Science Foundation.

Science and Technology Policy:
Perspectives for the 1980s

Preface

THE CONTENTS OF THIS VOLUME are based upon papers and summaries of discussions presented at a Conference that brought together leading figures from the developed countries of the world to review a wide range of policy issues in science and technology. The Conference was intended to provide a pause in the midst of increasingly hectic national and international activity to place in better perspective decisions on the use of science and technology in our society and the ways in which these decisions are implemented.

The unique quality of the Conference derived from the participants—approximately 75 individuals from the United States and Europe whose renowned expertise and professional responsibilities in government, industry, and universities place them in a position of extensive involvement in science and technology policy issues. The experience and judgments of these individuals therefore provided a rare forum of opinion on three broad questions:

1. What are the current major issues in science and technology policy?

2. What appear to be the trends—what new problems are anticipated within the next five years and what new priorities may arise among the present ones?

3. Given the growing experiences and experiments of different countries, what approaches to these issues appear to be effective, and to be recommended for emphasis, for both government and the private sector?

More specifically, the Conference focused on the relationship of science and technology (1) to such national concerns as economic growth, the impact of regulation, and foreign policy and (2) to technical fields including energy, materials, and communications.

This volume contains the formal papers presented at the plenary sessions of the Conference. In addition, it contains the papers presented at each panel session to stimulate and focus discussion, as well as summaries of these discussions and comments prepared by panel chairmen. Inasmuch as this volume considers current and near-term problems and recommended approaches to solutions, it is intended to provide a reference point for western industrialized countries to develop policy guidelines for improving the health of science and the usefulness of technology. As a specific reference point, the volume is designed to assist the National Science Foundation in its preparation of the Five Year Outlook Report.

It is clear that there is no single, all-inclusive "science and technology policy." The interaction of technical developments within our society is too varied and pervasive for any neat formula. Nevertheless, it is useful to identify and assess the common threads that are interwoven in such a complex tapestry. It is hoped that this volume will provide insights that may have generalized application for evaluating many aspects of science and technology problems, the roles of public and private sector actions, the relationships among the many institutions involved, and those policies that can address most effectively the problems that we now confront.

We are very grateful for the funding provided by the National Science Foundation to support the Conference held in March 1979 that gave rise to this volume. This financial assistance afforded us the opportunity to broaden the scope of the issues addressed and to increase the level of participation. We wish to express our deepest thanks for the Foundation's support.

We are especially indebted to the Conferees for enriching the Conference by their participation. In this regard, special thanks are due to our colleagues from abroad. The individuals who made presentations at the Conference and worked with us to prepare their papers for publication also deserve special acknowledgment of our gratitude. The rapporteurs competently and promptly assisted the panel chairmen in preparing drafts of discussions and their participation was essential to these proceedings. Throughout this project, several people provided exceptional support, and we wish to express our thanks to Mary Damask, Neal Brodsky, Kate Zadek, and especially Barbara Muench.

—HERBERT I. FUSFELD
—CARMELA S. HAKLISCH

Overview of Science and Technology Policy—1979

HERBERT I. FUSFELD

Objectives

It is the intent of this paper to provide an interpretive view of the nature, the range, and the trends of science and technology policy today. Many specific topics are discussed in some depth in the papers contained in this *Annal*. It is hoped that this paper will identify some of the common threads that are present in the varied range of subject matters discussed, and will establish a framework for the current state of science and technology policy into which these separate contributions can be placed.

A summary of the international Conference from which this volume evolved is given in the final paper by W. D. Carey. This paper provides, from Carey's unique perspective, a sense of the major themes and overall impressions arising from the two days of presentations and discussions that made up the Conference.

What is Science and Technology Policy?

There is often an elusive and unstructured quality associated with the concept of a national or international "science and technology policy." Nevertheless, while it is often the focus of scholarly debate and research, this concept bears directly on the expenditures in 1978 of about $48 billion in the United States alone, affecting the efforts of about 600,000 professional scientists and engineers.

This paradox of an apparently esoteric subject guiding and controlling highly tangible, critical activities is more semantic than actual. There is a very real and focused role for science and technology in materials, energy, and communications. There is direct involvement of science and technology in more general areas as economic growth, foreign policy, and regulatory objectives. In each case there are well-defined issues con-

Herbert I. Fusfeld is Director of the Center for Science and Technology Policy at New York University, former Director of Research for the Kennecott Copper Corporation, and a former President of the Industrial Research Institute.

1

0077-8923/79/0334-0001$01.75/2 © 1979, NYAS

cerning the allocation of technical manpower and resources, and the specific goals to which science and technology can contribute.

The sum of these discrete packages makes up our "science and technology policy." It is a description, not a focused objective in itself. The issues and policies within the separate packages are not necessarily related and they may not necessarily be consistent. Hence, the orderliness and relatively sharp definitions that exist for each area do not produce an equally ordered and sharp view of the resulting assembly.

The issues which, collectively, constitute science and technology policy can be considered in two broad categories:

1. *Mechanisms by which science and technology can contribute to solutions of particular problems of society and the economy.* Each problem area—whether as specific as energy or as general as economic growth—will require unique mechanisms for involving science and technology. The policies that create or stimulate these mechanisms can only be discussed in terms of the area, for example, science and technology policy in materials or in foreign policy.

2. *Mechanisms for strengthening the infrastructure of science and technology.* Each of the industrialized nations builds its technical competence upon: (1) a supply of trained scientists and engineers, and thus on the educational system responsible for training; (2) the reservoir of basic science and engineering available, particularly those fields to which their own national resources contribute; and (3) the interactions among broad sectors of interest—government, university, and industry—which permit the most effective use of national resources.

The papers contained in this *Annal* focus primarily on the first category. However, some aspects of training, of the continued need for a healthy structure of basic science, and the necessity for cooperative efforts are touched on in other papers in this volume.

Current Concerns of Science and Technology Policy

There is clearly a sense of malaise in the Western industrialized nations during this past decade, and the focus on science and technology policy today is simply one manifestation of this feeling. In the simplest terms, three changes have occurred in our economic and political environment which account for many of the current issues that make up our science and technology policy:

1. There is now public focus on contributions expected from science and technology to the solution of certain broad societal problems such as economic growth and foreign policy.

2. Significant changes have occurred within each national environment in several important factors (e.g., regulatory policies, societal attitudes, and investment climate) which influence strongly the private decisions that must be made as to the direction, timing, and level of industrial research and development (R&D) efforts.

3. There have been critical changes in the international environment that affect the conditions for world trade and the availability of energy and raw materials.

Our perception of the problems that arise from these changes and, more importantly, of the approaches to be considered for solutions will be conditioned by our interpretation of the societal trends that underlie them. There are many who believe that the convergence of economic, political and social changes is coincidental or, at worst, cyclical, and that the nature of tomorrow's society and economy will not be greatly changed. To these "conjuncturalists," any mechanisms to help solve today's problems must be compatible with relatively constant economic and political behavior.

Others, probably the majority, believe that the problems emerging from both national and international sources represent fundamental changes in the foundation of our society and economy. Hence, for these "structuralists," solutions to the problems of today must be geared to the society of tomorrow.

The challenge to those concerned with science and technology policy is twofold. First, the time span of 10 to 20 years is probably too short to permit conclusive evidence as to which societal changes are fundamental and lasting, so that our actions will be guided as much by conviction as by facts. Second, the choices made by today's society concerning the allocation of our technical resources will be a factor in *determining* the society of tomorrow.

It is therefore particularly critical that, in our search for solutions, we consider the relations between the issues for science and technology policy and the broader economic or political issues with which they interact. These interactions underlie the comments that follow for each of the three areas of change identified at the beginning of this section.

Focus on Expected Technical Contributions to Societal Problems

There is today a general acceptance, whether through conviction or resignation, that there will be increasing actions by government to bring science and technology to bear on problems in the "civilian sector," that is, in areas other than those of the military and space programs. This

The Western industrialized nations as well as Japan are experiencing problems in differing degree that relate to the economic and social changes of the past twenty years. More precisely, although the underlying causes of these problems can be traced back to factors throughout our common industrial growth of a century or more, there are a number of specific problems of an increasingly critical nature that are evident today. This statement could surely have been made at any period in our history. The unique feature today is the impact of certain key problems on our allocation of technical resources, particularly on the conduct and direction of industrial research.

Perhaps the most critical feature of our political-economic environment today is the pervasive sense of regulation that affects our industrial processes and products. This creates important issues for science and technology policy.

In the simplest terms, regulatory actions become a strong force for allocating and directing our technical resources. To the extent that this is an activity by which science and technology can help in the achievement of social objectives, it becomes one of the important and useful roles of government. To the extent that it takes technical resources from other desirable objectives and limits the flexibility for the optimal social use of science and technology, it becomes an intrusion by government in the processes by which a democratic society balances the use of its total resources to meet a large number of social objectives, not all amenable to government actions.

Specifically, regulations impact directly and indirectly on our technical efforts. The setting of specifications for health and safety in products and processes helps to create objectives for research and development programs. This can be a stimulant for activity by the manufacturer to develop substitute processes and to redesign products, and for suppliers to develop control equipment, improved material properties, and so on. This produces additional sales for suppliers but there is normally an added cost to produce the same sales for the purchaser. Thus, for the most part, this represents a trade-off between a social objective and economic growth.

This direct trade-off is at least definable, although acceptable methods must still be developed to put an economic value on a social objective. Nevertheless, it provides some guideline for society in deciding the proper extent of regulations as far as these more obvious economic impacts are concerned. It is far more difficult to evaluate the principal indirect impacts, such as the following:

1. Given a finite reservoir of technical resources in the country, and a

finite R&D budget for a given company, the technical effort devoted to meeting regulatory objectives leaves less technical effort available for other objectives. In particular, traditional industrial research intended to yield new products or processes would be less than it would be if *all* corporate R&D were devoted to these purposes. Thus, there will be a decreased basis for overall economic growth that must surely be more significant than the new products developed specifically for regulatory controls.

2. The process changes or redesign and retooling changes necessary to comply with the physical or procedural specifications of safety and health regulations can require very large investments, especially for the heavy, capital-intensive industries. This can result in: (1) causing certain product lines or business areas to become marginal as candidates for new investments, thereby eliminating the justification for continuing R&D in those areas; and (2) lowering the capital available for either expansion and improvement of present businesses or diversification into new businesses. When this reduction is significant, as in the mining industry, it can lead to a cutback in R&D, since the results could not be exploited by the company.

3. Most subjective, least tangible, but surely critical is the psychological impact on decision-making for corporate investments in change and growth, and hence in corporate support for longer-range or innovative R&D that could lead to substantive changes in business direction or product lines. There is a fundamental climate of uncertainty that accompanies a climate of regulation. This is related to the uncertainty in actions that are controlled by people and by political processes. There has indeed been a history of change in specifications and in public philosophy regarding the nature of health and safety as they relate to industrial products and processes.

Time, the necessary ally of research, is almost inevitably the natural enemy of investment. Time, when coupled with inflation, can be enervating and destructive to R&D investment. Time, which includes a high probability of adverse change (societal as well as economic or technical), can be fatal to such investment.

It is the sum of these several indirect impacts of regulatory actions that provides a focus for science and technology policies. Every difficulty identified becomes an issue for correction. Hence, we become concerned with the need to provide a scientific base for regulatory actions and thus to minimize arbitrariness and inconsistencies. And there is intensive preoccupation with mechanisms that will reduce private risks or increase rewards, so that the ratio will induce private investment in R&D. The major directions of such actions tend to be: (1) tax or investment policies of government to favor increased private R&D or increased investment in its application; and (2) direct government cost-sharing or support of those phases of R&D appropriate to the industry involved.

The appropriate actions differ in each sector, as the papers in each of the specific technical areas demonstrate. There is considerable room for misunderstanding between government and industry, which is shown clearly in the paper by L. Branscomb, which considers the current attempts by the Department of Transportation to find a role for government in development of the automobile of the future.

There is an implied proposition running through considerations of each subject related to civilian-sector R&D, and touched on more generally by A. M. Bueche and W. D. Carey, which can be described in the following way:

A set of economic, political, and social changes has made R&D in industrialized countries a less-favored investment choice for the private sector than it was in the 1960s. This is due in great part to a sense of uncertainty about the critical considerations on which the investor must rely. Government can take some action to minimize the uncertainty and improve the investment climate, but to do this effectively requires the involvement of the private sector in planning, cooperation in R&D, and application of results. And this, in turn, requires public acceptance of two principles: (1) that the priorities and judgments of the private sector be accepted as a major planning element in economic growth, and (2) that public funds may create a private profit in the process of accomplishing national objectives.

Thus, effective science and technology policies that address the new national environments of the Western industrialized countries quickly become a part of political and economic policies. This is both reasonable and inevitable, and simply re-emphasizes the intertwining of science and technology policy with general public policy. The comments of D. Davies offer a broader perspective on these interactions.

Changes in International Environment Affecting R&D

Nowadays the ties that bind and the barriers that divide often appear, on closer examination, to have a common root in the growth of modern science and technology. Not that our technical growth inevitably creates ties and barriers, but simply that nations can always find new tools that serve to unite them, or new criteria for dividing the "haves" and "have-nots."

Because we seem, as people, to have the capacity for building friendships or antagonisms from any set of conditions, it is hardly surprising that the changing international environment becomes the source of several critical issues for science and technology policy, that is, mechanisms and principles by which science and technology can contribute to solving problems of an international character. In brief, there are opportunities, and perhaps imperatives, for science and technology to become tools for national foreign policy in concert with traditional economic and political approaches.

Out of the enormous range of international concerns that surround us all, the broad categories that identify the key technical efforts appear to be:

1. Cooperation between developed and developing countries with primary emphasis on the technical-economic growth requirements of the latter.

2. Technology exchanges between OECD countries and those with centrally planned economies.

3. Development of equilibrium in international trade, primarily among developed countries.

4. Needs of developed countries for raw materials and fuels.

The first three are conveniently geographical, giving rise to the commonly used simplification of North-South, East-West, and West-West. The fourth is largely technical, although intensely political partly because of the geographic spread of natural resources, and partly because of the sense of difference in lifestyles and per capita income between producing countries in the developing world and the industrialized nations.

Elements of these concerns have been present, in different forms, throughout modern history and undoubtedly long before that. The categories could be looked on alternatively as:

1. Trade between haves and have-nots, between the advanced industrial and/or commercial regions and the primarily agricultural, nonindustrialized regions.

2. Trade between past, present, or potential antagonists.

3. Trade between friendly nations.

4. Actions of countries whose manufacturers depend on raw-material imports.

In other words, there is no need to recite a litany of international economic and political considerations that our historical perception tells us have been present—with different actors, locales, and subjects—in other eras. We should, instead, focus on the common threads in the international tapestry that are unique to the technical nature of the present.

There is one principal component in these considerations, not necessarily the most critical for any given circumstance, but surely the source of much misunderstanding between nations and within each nation. This is the fact that, while governments negotiate and set conditions for international trade and conduct, much of the modern science and technology that stimulates economic growth and improves living standards belongs to the private companies of the countries of the Organization for Economic Cooperation and Development (OECD).

The source of misunderstanding is that we are not talking about "packages" of science and technology that can be taken from their private owners and simply transferred to another country where they will have equal value and usefulness. The "ownership" consists of a broad system which pursues technology in parallel with capabilities for manufacture and distribution, then manages the integration of new products and processes with these capabilities so that the overall system is in balance.

We are, in short, referring not so much to "technology transfer" as we are to "system transfer." This involves a body of principles, organizational structures, and even psychological motivations that are not within the traditional competence or authority of Western governments to sell or transfer.

This statement can be put in a more constructive form. In each of the international areas of concern mentioned at the beginning of this section, the developed countries of the OECD can pursue their foreign policy goals, or perhaps obligations, most effectively through the cooperative involvement of the private sectors in those countries. While this is probably true in almost every area of international concern, it is uniquely true in those to which science and technology can contribute, particularly because of the modern evolution of industrial research. It is this evolution that has allowed the development of methodologies for the planned integration of R&D into the broader system of manufacture and distribution of today's industrial society.

The theme of reconciling the obligations and initiatives of government with the know-how and property of private companies defines many of the issues for science and technology policy that arise within the changing environment of international affairs. This is particularly important in the difficult task of considering the views, the objectives, and the conditions of developing countries.

These general comments can be made more specific by a brief discussion of each of the separate categories of international affairs referred to, that is, developing countries, exchanges with centrally planned economies, Western international trade, and supply of natural resources.

Cooperation with developing countries

Much of the government-to-government activity involves (1) public sector technology, such as that in health and agriculture, and (2) concern for the technical infrastructure of developing countries, by strengthening educational institutions and building a basic research capability, for example. This activity tends to be long range in nature. Shorter-term needs relating to employment and economic growth that can be helped by science and technology require the specific technologies plus managerial competence of private companies.

Clearly, a major issue is to determine mechanisms by which developing countries can work with the private sector of developed countries under conditions acceptable to both. A vital related question is the role of governments of developed countries in establishing such conditions and in facilitating the relationships that arise. Resolution of these issues should provide easier approaches to a third issue, namely, how can the strengthening of a scientific and technical infrastructure within a developing country be used effectively in the economic growth of that country?

It is obvious that these are not simple matters of trade, but rather of deep national pride and ambition, of widely different cultures and values, and of different economic and political structures. The problem becomes that of how to transfer competence in developing and using science and technology that evolved within the Western democratic and economic systems to nations that have chosen different approaches, often incompatible with those systems. The issues arising from this question quickly involve an intricate maze of politics, economics, and sociology mixed with the substantive questions of science and technology.

Fortunately, there are many examples to demonstrate that conditions can be established by which specific needs of developing countries can be met while permitting profitable operation of the private companies involved, and without requiring change in the political structure or objectives of the developing nation.

The complexity in accomplishing this on a general basis is reflected by the current difficulties in effecting a compromise on a "code of conduct" for such trade and technical exchanges. It may well be that both developed and developing countries must proceed by encouraging specific agreements and joint ventures whenever possible, developing an understanding of the criteria that must accompany those efforts which meet the minimal objectives of both parties, and evolve a reasonable "code of conduct" that can be joined by those countries who wish to do so.

This process can only succeed if there is a continuing effort to clarify the nature of "systems transfer" which permits the use of science and technology in economic growth. This requires examination of the relevant policy issues by institutions in both developed and developing countries, as well as the cooperative participation in such analyses by the organizations in the public and private sector that engage in the actual technical and industrial transactions.

A focus on these issues will surely be critical in the years ahead, since the relations between developed and developing countries touch on the next three areas of international concern, at least indirectly, and often very directly.

Technical exchanges with countries with centrally planned economies

While the high degree of planning in many of the industrialized countries in the OECD is recognized, the focus of this section is intended to be on socialist countries such as the USSR, the Eastern European countries, China, and Cuba. These countries all possess an industrial base and in terms of technical and industrial development are in a category quite different from that of developing countries. Nevertheless, there are certain factors in the East-West technical exchanges that have some similarities to the North-South exchanges.

Specifically, the civilian sectors of these planned-economy countries lag behind those of the West in technology and productivity. There is, further, a recognition that Western technology and research management can provide constructive inputs to these economies. Finally, there is a conscious government policy in those countries to encourage such exchanges.

This combination of factors has led to many examples of successful exchanges between a Western company and a centrally planned economy. These show clearly that the private sector of the West can operate profitably with countries of vastly different political and economic structures.

The overall pattern of exchanges in science and technology, coupled with the closely related industrial agreements involving plants, products, and licences, forms the basis for several key questions of science and technology policy. In all subjects involving international affairs, there is one underlying question: Can science and technology be useful tools in the conduct of foreign policy? In the case of the complex and ambivalent relations between OECD countries and those with centrally planned economies, there are several specific forms that this question takes in terms of the actions to which they can lead:

1. What is the value of science and technology exchanges to the countries with centrally planned economies and to the Western countries involved?

2. What is the appropriate role for Western governments to play regarding interactions between the technical communities of the West and those of the centrally planned economies?

3. By what mechanisms and under what conditions can or should the policies of Western governments be used to influence the actions by the private sector of the West regarding exchanges with the centrally planned economies?

These questions take on particular significance because of the political differences between the Western countries and those countries under consideration having a centrally planned economy. Thus, a major policy issue is the relationship between an improvement in some industrial sector resulting from a technical exchange and the military capability of the particular country with a centrally planned economy. This is a fundamental question far beyond the competence of this paper. It must be considered along with the byproduct benefits of technical exchanges for the overall political atmosphere, as well as the possible technical and economic benefits that may accrue to the Western countries.

The significance of this critical technical-economic-political entanglement is that actions may be called for that would not be appropriate in other areas of international affairs. Communication is a minimal requirement. Western governments must establish adequate mechanisms, formal or otherwise, for policy dialogues between government representatives and both the scientific community and the industrial sector.

More detailed understanding and analyses are required as to the technical and economic impacts of technical exchanges to all parties, public and private, Western and socialist. This calls for coherent research that involves technologists, economists, and specialists in the countries involved.

In the United States, the unique existence of antitrust laws raises special issues. The possible modification of these laws to permit collective industrial action in technical exchanges with countries with a centrally planned economy should be examined to analyze the impact on our domestic competitive situation as well as our competitive position in such exchanges.

There are, in brief, very great opportunities for political and economic benefits from these scientific and technological exchanges, and there is an equally great necessity to analyze carefully their impacts on the overall national policies involved.

Trade among developed countries

Since each nation's position is unique and varies with history, the changes in international environment can only be assessed by each nation separately. Thus, the remarks in this section are from a United States viewpoint today, although with some generalizations that can apply more universally.

The United States is faced with a serious balance-of-trade deficit which, while traceable in very great part to oil imports, has focused attention on a declining advantage in high-technology products and processes, and in technical innovation generally. This has also focused much attention on the government-industry mechanisms used by other industrialized countries to improve the world-competitive position of their industries. And there is a less direct, but undoubtedly critical, impact of the large multinational corporations, whose role in world trade and in technology transfer is intertwined with the questions of trade deficits, technical innovation, and government-industry mechanisms.

The relationship of trade to science and technology is an economic question. There is nothing to be added here to the straightforward proposition that whatever science and technology can do to increase productivity, lower costs, develop new products and processes, and increase receipts from technical licensing, will strengthen the position of the United States. Hence, those science and technology policies that address domestic economic concerns bear almost as directly on international trade.

There are, nevertheless, several aspects of international trade that involve science and technology beyond the domestic situation. Among these are the following:

1. Regulatory actions in such areas as safety and health are proper and necessary functions of government. They can, however, add substantially to the financial requirements of particular industries, and thereby become an additional barrier to domestic innovation. They can become a cost factor that places certain products at a competitive disadvantage internationally. This places great importance on reaching international agreements among the OECD countries at least, and preferably with all countries, to agree on common standards for health and safety in final product performance and in manufacturing. This in turn will emphasize the need to provide a reasonable scientific basis for regulation that will be consistent with international views of cost-benefit analysis.

2. Much attention has focused on the impact of tax laws on the level of industrial research. There is concern over one interpretation that would not permit an American multinational corporation to deduct its domestic research costs from its domestic sales, but would attempt to allocate the proportional part of this research to foreign sales. Since this would not be accepted as a proper expense by other countries, the company would not receive full credit for its research. This provides a financial incentive for United States companies with foreign operations to conduct more of their research in foreign countries where the research costs could be deducted from sales in those countries. The effect could be to decrease further United States generation of technology and worsen our balance of trade deficit.

3. In considering the difficulties of the United States, we are led to study the actions that other governments take to improve the international exports of their industries. A principal feature appears to be a closer relationship among planning, financing, and business affairs. This is not necessarily a friendlier relationship, except perhaps in Japan, and it is yet unproved that it is a more effective relationship than the adversary relationship within the United States. The overall impression from the complex of financial subsidies, cooperation among companies encouraged or stimulated by government, and the attempts to combine national objectives with the industrial research programs of private companies is that the United States would have to consider two guidelines that would begin to match these efforts: (a) As referred to earlier in the discussion on the effects of changing national environments, we would have to permit private profit to be derived from the use of public funds, although under conditions of achieving specified national objectives for the technical efforts involved, and (b) we would have to consider whether our antitrust laws could be so modified that companies could act collectively to meet an international objective while maintaining the objectives of competition domestically.

Thus, science and technology policies that affect our international trade again focus on the interaction of government with the private sector. Progress, or at least change, seems to depend as much or more on legislation as on technical developments. The role of government is clearly to choose wisely regarding the first in order to encourage and utilize the second.

Needs for natural resources

The dramatic changes in the availability of oil, and the sudden awareness that access to other raw materials could follow a similar course, make certain issues for science and technology policy almost too obvious, for example, development of alternate sources. But even the obvious becomes complex on a global scale.

Obtaining natural resources from another country is hardly a new phenomenon. A particular feature of today's environment is the concentrated structure of modern industrialized society, its dependence on high-grade energy sources, and the intricacies of interdependence in an overall system of manufacture, distribution, and maintenance. One immediate characteristic of this system for science and technology policy is that technical changes will require long-term efforts, not only for the difficult technical developments in themselves, but also because of the time needed to introduce change into an ordered economic and technological structure with minimal disruption. Any short-term improvements can only be accomplished primarily through economic, political, and cultural changes, with technology providing modest immediate help, but holding out the promise of long-term support.

There is more flexibility and thus shorter-time potential for contributions in the nonfuel minerals area. There are often alternate sources of raw materials, substitutions possible in making present products, new design approaches that might perform a given function with a new product (and new materials), conservation possibilities in recycling and waste avoidance. Almost nothing can be changed immediately, but many changes in design, manufacture, and use are possible in a two- to five-year period, and very substantial changes in our materials situation can be effected in a five- to ten-year interval.

This all provides for a realistic and constructive role for science and technology policy, but calls again for the proper mix of government actions with the capabilities and business plans of the private sector, particularly in the mining and materials industry. The national objectives are clear. We want to minimize dependence on outside sources for materials that are important militarily or economically; we want to minimize the costs of new designs, substitute materials, and new domestic sources; and we want to conserve energy and maintain environmental quality.

We therefore require a trade-off in order to secure the necessary materials for our industrial society with minimal cost, minimal energy, and minimal pollution. Whenever there is a trade-off in specifications, there is a role for science and technology. This in turn defines the science and technology policy issues. There must be mechanisms to stimulate the needed R&D, public or private or both. There must be a willingness by society to permit an adequate ratio of rewards to investments so that the results of R&D will be applied to the extraction of materials from new sources and in the installation of new processes. There must be an awareness of the societal impacts with respect to energy and environment in the development of adequate natural resources, plus a sensitivity on the part of industry to the proper balance between economic requirements and quality of life. And there must be stimulation of interest in graduate schools so that the most promising future researchers become interested in these problems.

It is an interesting fact that different periods can be viewed from their technical characteristics. There are the classic Ages—Stone, Bronze, Iron. There are the energy periods—wood, coal, and petroleum. We witnessed the chemists' war in World War I and the physicists' war in World War II. In the materials field that followed World War II, the technical community has seen several such periods. There was first that of *materials science*, concerned with the understanding of atomic and electronic structures to develop materials with new properties needed for missiles, jet planes, nuclear energy, and solid-state electronics. This was followed by the period of *materials engineering*, emphasizing the most economic use of the right material in the right product design.

Today, our principal concern is with *materials supply*. Its needs and characteristics have been described. But our materials community has evolved from the relatively recent periods of materials science and materials engineering. There is a challenge to science and technology policies of the OECD countries to develop mechanisms that will shift the research and training priorities of the materials community to meet the new societal needs arising from the uncertain international politics that can disrupt our materials equilibrium.

Future Concerns for Science and Technology Policy

This final section will try to provide some perspective on the broad thrust of science and technology policy in the near future and thus will focus more on the environment in which these policies must operate rather than on the substance of each technical field.

Since we set a target of the next five years for looking ahead, many of the most likely areas of focus for science and technology policy are definable. While new specifics will arise, the desired changes in a society or economy, like manageable changes in our physical environment, relate tomorrow to today. Nature is rarely discontinuous. Hence, tomorrow's emphasis for science and technology policy arises from today's problems.

Transitions to New Lifestyles

Changes are not likely to be as dramatic as these words might imply. Nevertheless, as different problems of both national and international character are considered, it becomes clear that the objectives or the results of technical efforts will lead to less energy per capita, less materials per capita, and surely less pollution per capita.

On the surface, this statement may suggest the theme of "small is beautiful," or a call to "appropriate technology." But those phrases are most often used in connection with a nation of modest resources and income which is developing a society that can achieve satisfactory equilibrium with less energy and materials and lower cost than in the case of the industrialized nations to date. And the *usual* corollary is that this equilibrium is achieved within a less complex society.

These concepts do not apply to an industrialized nation with adequate resources and wealth and, above all, a strong base of science and technology. Less energy and materials per person? Yes. But simpler? Not at all. We look forward to a more sophisticated, more elevated society than can achieve its fundamental objectives and intellectual satisfactions by using less energy and materials through its advanced technical efforts.

We are active in communications advances, materials research, changes in manufacturing processes, creative systems of distribution, advances in transportation and, beyond these, possible shifts in city planning. Does "less" mean "simpler" or "inferior?" Compare the lighter-weight, fuel-saving car of today with one of just five years ago. The efficiencies of solid-state ignition and fuel injection, the steady improvement of strength-to-weight ratio in materials, and the safety features in design, all represent a technically sophisticated shift to a less energy-intensive and materials-intensive society. And these tangible examples are dwarfed by the incredible advances created by microprocessors in communications and controls that can make our home and work routines more efficient and hence less energy consuming.

A common issue for science and technology policy in many fields will follow this principle. How can one realistically carry on the functions of society with less energy and less materials? What incentives, what government role, what cooperative mechanisms are helpful in each specific area to accomplish this? The goal will not be how to live more simply, but more intelligently, and the part played by science and technology in its achievement will be critical.

Productivity of Science and Technology

In an idealized view of the growth of modern society, the conduct of research and the development of new technology were carried out formerly in parallel with, but not as an integral part of, the daily functioning of society. That is, while society received with pleasure, profit, and amazement the constant stream of wonders emerging from the technical community, society proceeded with its daily chores and future plans without expecting or requiring technical change. Thus, apart from deciding what fraction of its resources to put aside for science and technology, society could leave that community alone until particular results required attention.

If such an idealized time ever existed, it is surely not now, when our future economic plans, our concerns with resources, our very quality of existence in any degree of peace and comfort all appear to require contributions from science and technology. Society not only needs these contributions, but it *recognizes* these needs and therefore it *expects* productivity of each national technical enterprise in all aspects.

"Productivity" is a word that must be carefully used in connection with R&D. It has the same general meaning as in manufacturing and other areas of economic activity, but the specifics differ. Generally, increasing productivity implies that we get more "results" for a given cost or manpower level, or that a desired "output" or "result" can be obtained at a lower cost or reduced manpower level.

Obviously, the key lies in our notion of "results." Purely quantitative measures of such direct outputs as patents or publications are very crude indicators of *activity*. But in the expectations of society, or of scientists, activity is only a means to an end. Society wants new products, processes, and procedures, and scientists want new understanding, new concepts, and new frontiers.

There are two categories of results that should offer a more complete definition of productivity in technical activities: (a) results should have some impact on our wellbeing, either in the form of new goods or lower costs from the private sector, or as new services and community benefits from the public sector, and (b) results should provide definable advances in our reservoir of basic science and engineering.

The evaluation of the output of basic research or of goods and services resulting from R&D is very largely a matter of judgment, but a number of quantitative measures are available. An individual company can certainly measure, with much argument, the value of its internal R&D. Other measures can be devised to evaluate the "social value" of R&D to the community at large, as suggested in the work of E. Mansfield and R.R. Nelson.

For the purposes of this discussion, let us agree that there is a productivity of science and technology in both private and public sectors, and that it can be measured or judged or both. Because the costs of R&D are so great (about $50 billion in the United States today) and the functions it serves are so important to society, there are substantial tangible benefits to be gained from improving our technical productivity.

The mechanisms for such improvement will be important matters for science and technology policy in the next several years. They will include the following items:

1. There is a classic analogy with conventional manufacturing productivity in that investment in equipment and facilities can increase the quantity and nature of technical output. A linear accelerator or radio telescope represent large investments to permit advances in basic science. Automated analytical equipment, computers for calculation or for storage and retrieval, improved methods for control and automation provide more and different laboratory outputs with a given level of manpower.

This consideration leads to a simple type of policy question. What fraction of available funds should go to facilities and equipment and what to manpower? Certain fields are more capital-intensive than others, so that expansion of funds will not account for proportionate expansion in the number of scientists. Our expectations of the nature of results and of time scales must be compatible with these decisions.

2. Since we measure the productivity of the total technical enterprise in terms of its impact on our wellbeing, we must look at the total system with an eye to potential improvement. Two aspects of this issue that are receiving attention and that will be highly emphasized in the next few years, are: (a) How do we select the right programs to work on? (b) How can we smooth the transition from R&D results to manufacture, distribution, and use?

The answers to both questions require the involvement of the private sector. The critical challenge for science and technology policy is to see that we devise mechanisms by which inputs from the private sector are available in planning all technical programs, public and private; that we create incentives and eliminate barriers in transferring R&D results from public control to private initiatives and utilization when that is appropriate; and that we establish conditions to expedite private exploitation of private sector R&D.

3. The three principal parts of our technical community are government, universities, and private industry. Each sector has a vital role in our society and thus the nature of technical activity in each has a different character. In carrying out these separate technical activities, we recognize that: (a) Each sector has competence and/or specific knowledge that can help the others in planning R&D and in avoiding inadvertent duplication. (b) A technical program or an objective of any one sector can often draw upon the efforts or special skills of the others, and thus accomplish more with less in a shorter time.

Hence, a principal focus of science and technology policy will be to devise specific mechanisms for cooperation among these three sectors, including incentives that will encourage voluntary cooperation.

The anticipated emphasis on productivity is not a matter of bureaucratic niceties, but of the deeply-felt belief that the nature of today's societal problems *can* be helped by science and technology and even require their involvement. This presents a problem to the administrators and policy-makers of technical institutions to present for public understanding the variety of their technical activities. Nevertheless, a rising public faith that our technical resources are being used effectively for society's purposes (if that is what we mean by productivity) will provide stability and support for the scientific and technological enterprise of each nation.

Expanded Cooperation with Developing Countries

This will clearly be a major preoccupation for science and technology policy in the future. The United Nations Conference on Science and Technology in Development (UNCSTD) takes place in August 1979 in Vienna. It culminates years of effort, especially intense in the past year, and will initiate actions to show that the industrialized nations are sincere in establishing programs of technical cooperation.

Sincerity is not enough. The years following UNCSTD will test whether the industrialized nations and the developing nations can work out mutually acceptable objectives and mechanisms to carry out these programs. Each program of technical cooperation must specify the period of time needed to have an impact so that expectations of results are balanced and understood in advance.

The establishment of an Institute for Scientific and Technological Cooperation (ISTC) by the United States will provide a unique opportunity for American science and technology policy. What priorities should be assigned for developing countries and what interactions can ISTC establish with them? The satisfactory operation of this Institute could well provide an example for other industrialized countries to follow, possibly leading to a broader institution some day involving a number of these nations.

A "code of conduct" for technology transfer is a question that has not been resolved, and is unlikely to be in the near future. This issue must be addressed again after the Vienna Conference. A major area for policy study will be to analyze the outcome of the Conference, identify the principal activities to be pursued by the developed countries in accordance with the objectives and receptivity articulated by the developing countries, and then to re-evaluate the need for a formal international code of conduct and its scope in view of the climate following the Conference.

The principal issue for science and technology policy will continue to be that of the role of government in negotiating cooperative agreements that involve the competence and property of the Western private sector. The results of UNCSTD should provide suggested directions, but the solution must be determined by each national government separately.

Emphasis on Natural Resources

The difficulties with balancing our need for fuels and minerals against available supplies will be high on our technical agenda, and therefore a major focus for science and technology policy.

There are important and varied technical programs related to production, consumption, and substitution in the area of natural resources. However, all that science and technology can do is to offer options for the solution of problems. The actual integration of change in our system involves economic and political forces, and any short-term change depends almost entirely on such forces.

The factors that affect the use of natural resources, apart from physical and chemical properties, are their cost, their availability, and their location. The problems posed by each factor can be addressed by specific technical programs. The principal issue posed for science and technology policy will be to select the emphasis for technical priorities that will be compatible with overall economic and political strategies. This is generally true for any area of interest, but it is of unique importance for natural resources because of their intimate entanglement with economics, politics, and society. The paper by Lord Rothschild indicates one technical approach that may minimize the economic and political obstacles, but they are present nonetheless.

As an example, there is the continuing dilemma of what to do about ocean mining so that jurisdictions may be established to secure and process manganese nodules with minimal disruption to the ocean environment and to enter their products into world commerce at competitive cost. Currently, incentives to stimulate private investment are marginal in view of uncertainties in international agreements as well as the cost of new technologies. Is there an appropriate role for government in aiding the required technical developments so as to minimize investment risks? And is it compatible with the government's position in the international arena regarding ownership and control? Furthermore, the United States is concerned because almost 100 percent of its manganese, a vital alloying element in steel, is now imported. Should we pursue technical programs to develop new matallurgical processes to extract manganese from high-cost alternative ores, or should we consider offering a modest guaranteed price for manganese produced from ocean nodules, thus making ocean mining attractive economically? And what impact would these measures have on the countries that now provide the manganese?

These types of technical-economic-political considerations are present in all natural-resource difficulties. Umberto Colombo, chairman of the Italian Atomic Energy Commission, has stated that "oil is not a scarce resource, it is a vulnerable resource." Vulnerability and substitutability obviously set different values on the scientific and technical priorities to be pursued. They change the conventional economic cost-benefit analyses for such technologies as coal gasification and producing oil from shale.

Science and technology policy in this field within the United States will require the cooperative involvement of the Department of State, Commerce, Interior, and Energy, as well as technical agencies such as the National Aeronautic and Space Administration (NASA), and the National Science Foundation (NSF), plus the active cooperation of the private sector. It is an excellent example of the complexities of public policies, and may therefore prove the most difficult challenge for our technical enterprise.

Conflicts in Regulatory Objectives

Since this problem is currently reaching a boiling point of political and economic frustration, the search for solutions will surely occupy science and technology during the next five years. Although the problems and solutions are vastly complex, the future role of science and technology can at least be stated simply.

Western industrialized societies have set forth a number of worthwhile objectives. We wish to minimize energy use and prevent environmental pollution. We would like to have safe products and improve safety in the work place. And we wish to reduce inflation and promote reasonable economic growth.

There are two features of these objectives that define our concern today:

1. In the technically based industrialized nations, these objectives conflict, and conflict sharply. It is not a conflict of government versus industry, or of the public interest versus private interest, but a conflict of public objective versus public objective—a fundamental, physical, inevitable conflict of a pluralistic society. We can do our best, we can compromise, we can make a little progress in each area, but we cannot in fact pursue each objective to its extreme and absolute completion without violent damage to other objectives.

2. Government has taken on the role of spelling out the goals, the specifications, and often the procedures by which the goals of each separate national objective are to be achieved. However, despite appropriate caveats in each law and regulation, the legislative treatment and the executive pursuit of each national objective is conducted as if this were the only national objective. The agencies responsible for developing the mechanisms to achieve these objectives often act as if they must indeed pursue them to their extreme and absolute completion.

As these objectives come into conflict, we attempt to find trade-offs. One important function of science and technology is to provide new options for such trade-offs, permitting us to go farther toward accommodating two or more presently competing objectives than would be possible with existing technology. Can we develop new engines or fuels that can reduce pollution while decreasing oil consumption? Can new metallurgical processes be developed to lower smelter emissions without increasing energy use? Can we reduce all health, safety, and pollution risks in any given industrial activity without making that activity uneconomic or perhaps inoperable?

The two functions of science and technology in these concerns are to reduce uncertainty and to provide new options for tradeoffs. The two functions of science and technology *policy* are to develop public understanding and encouragement of these contributions, and to devise mechanisms and incentives that will direct technical activity to these problems and expedite the use of their results to benefit society.

Regulatory activities in the broadest sense attempt to minimize risk—the risk of being hurt, of running out of energy, of suffering economic harm, of being cheated, and so on. In areas involving physical actions, there is a critical role for science and technology in providing a basis for separating unacceptable risks from acceptable risks, despite the deeply subjective nature of these terms. The more flexibility we achieve by providing a sound scientific basis for regulatory actions, the more rational the trade-offs. There is no complete removal of risk, and thus we can never do more than compromise our conflicting objectives. But science and technology can make possible reasonable compromise.

David Swan has used a phrase, "technology is the only free lunch around." When we have identified conflicts and risks in our objectives, we have thereby defined problem areas for science and technology and a role for science and technology policy.

Concluding Remarks

Many points have been touched on in this overview, but many have been omitted. I have chosen not to comment on specific technical areas of science and technology policy, except for natural resources, but have attempted to prepare and present a framework for classifying groups of issues and for identifying the key questions involved in each grouping.

I have referred in many instances to the fact that effective science and technology policy often requires interaction between the public and private sectors, since so many of the difficult problems facing us today are largely economic in nature. And in our form of society, the private sector offers a powerful mechanism for allocating and balancing resources to convert technology to use within the limits set by reasonable economic criteria.

Clearly, the economic objectives of society as a whole are far broader than the economic objectives served by the private sector alone. There are, additionally, political and social objectives that are not easily quantifiable in economic terms, so that judgments on priorities and use of resources involve more than the private sector.

Nevertheless, the majority of issues involving science and technology policy call for the setting of priorities among areas of R&D, arrangements for the efficient conduct of R&D, and mechanisms for integrating the results of R&D into our economic lives. These are all activities in which the skills and methodologies of modern industrial research are applicable, and for which the facilities, investments, and management of the private sector are necessary.

We must ask, therefore, how government can work with the private sector without unduly influencing the first or weakening the second. Fortunately, in the areas of science and technology, feasible mechanisms for specific goals appear possible without fundamental changes in structure.

Our adversary system is a useful set of checks and balances, as pointed out in the panel report by W. Dill. But reasonable modifications of this system, whenever general progress towards societal goals appear hopelessly entangled in conflicting objectives, would seem to be a small price to pay for vastly improved effectiveness. These could include more flexibility written into legislation and regulations, some forms of voluntary referee system before resorting to the courts, and a mutually agreeable scientific foundation in areas of controversy. It requires a common-sense view on the part of both government and industry as to the dependence of one on the other.

There will, and must, be much emphasis in the near future on these public/private interactions in regulatory activities affecting R&D, in university/industry cooperation, in resolving dilemmas related to the use of natural resources, and in the role of technology in trade and foreign policy. This raises broad questions of political and economic philosophy that apply to all phases of our society and form of government. The appropriate forum for resolving such questions in our democratic society is the political arena.

The questions raised by the conflicting approaches of government and of the private sector may seem too complex to present as falling under a general principle, so that no one guideline or piece of legislation can possibly resolve the overall issue of public/private interactions. Perhaps in the single area of science and technology policy, we can make some progress toward balancing multiple interests and developing pragmatic, workable mechanisms that can serve as examples for solving other problem areas in our society.

Back to the Bio-Board

LORD ROTHSCHILD

IT SEEMS SOMEHOW ANOMALOUS—almost ridiculous—for this paper to have been prepared by someone who has spent the twenty-five best scientific years of his life only examining spermatozoa before they collide with an egg, and then keeping an eye on them for a few minutes after the collision has taken place. But these studies have led me into strange fields. How can some sperm tell the difference between the isomers maleic and fumaric acid?[1] Why do they swim upstream in a parabolic velocity gradient?[2] Have they got sense organs? If so, where are they? How do they feel in General Electric's adiabatic calorimeter in Schenectady, where I put them for about a fortnight, much to *my* pleasure? How do they stir the medium in which they swim, given that hydrodynamics tells us that they can not?[3] How do they tell eggs to stop fooling about and start synthesizing DNA? Needless to say these questions were too much for me; so when the Shell Company cast a rather tricky fly over me, I swallowed the bait and became a research administrator. Research in Shell, in spite of having some unexpected similarities with my spermatologic extravagances, inevitably turned me into an energy man. So, ever since I left my laboratory in Cambridge, England to join Shell, consciously or subconsciously, energy in general and crude oil in particular have been on my mind. So I will say a few unoriginal words on this subject—how we trap energy and what we do with it.

It is doubtless a consequence of those years spent with oscillating rods in a viscous medium, as the hydrodynamagicians call them, that I have a predilection for specific issues. In spite of that, I believe that the most constructive way to tackle the problems of science policy and national

---

Lord Rothschild is former Director General of the Central Policy Review Staff, United Kingdom, and is currently Chairman of Rothschilds Continuation Ltd.

0077-8923/79/0334-0027$01.75/2 © 1979, NYAS

needs is to dilute such lofty considerations with specific practical issues. I propose to do this here by looking at a field of infinite diversity in which economic and political forces are likely to impose quite new and exacting demands on science and technology.

Insofar as oil is going to become more and more expensive, in real as well as in current terms, and since it is certainly going to run out or cease to be of major importance at some indeterminate time in the future, it seems to me of secondary interest whether Exxon, Walter Levy or Peter Odell is right about when these things are going to happen, assuming that these authorities have pronounced on this matter. What interests me about the subject is this: there are many who worry about the effect of future oil scarcity on what might be called our energy transformation economy, whether it concerns coal, shale, wave power, fusion, breeders, serious conservation (non-existent at present) or the sun. But how many of those who are known as "the man or woman in the street" are worried about or even aware of the almost incredible pervasiveness of oil in the non-thermal parts of our lives. Would 50 percent of a random sample of the people in this country know what the objects listed in TABLE 1 have in common? A sizeable fraction of each of these products is made from crude oil by way of petrochemicals. FIGURE 1 shows how the materials from which these products are made are obtained from crude oil.

Petrochemicals are made in a way that is somewhat reminiscent of genetic engineering. In both cases large, complex molecules are cut up into smaller pieces: in the one case by submicroscopic scissors (particular enzymes), in the other by giant plants such as ethylene crackers. In both cases, the smaller pieces are then stuck together again in a different way.

TABLE 1

PRODUCTS USING OIL IN THEIR MANUFACTURE

Panties	Medical syringes
Glue	Cameras
Hose-pipes	Egg boxes
Wigs	Fertilizers
Gear wheels	Telephones
Printing ink	Candles
Television sets	Sponges
Bubble-baths	Paint
Lipstick	Railings
Safety helmets	Lavatory seats
Shoes	Tires
Bottles	Sweaters
Venetian blinds	Greenhouses

In the case of crude oil the smaller pieces are the building blocks from which other, more complex materials are made. Glycerol, for example, may be derived from acrolein, which in turn can be made from propylene by partial oxidation. The ubiquitous polyethylene is derived from ethylene by polymerization, and the same applies to many other petrochemicals whose energy content is characteristically as little as one-third of that in the crude oil used in their manufacture. This energy loss was unimportant until recently because oil was cheap. Indeed, the low price of crude oil in the United States contributed to the exclusion of foreign petrochemicals.

Because of low world prices in the 1950s and 1960s, oil became the major source of energy and chemicals in many countries. Low-cost oil-derived

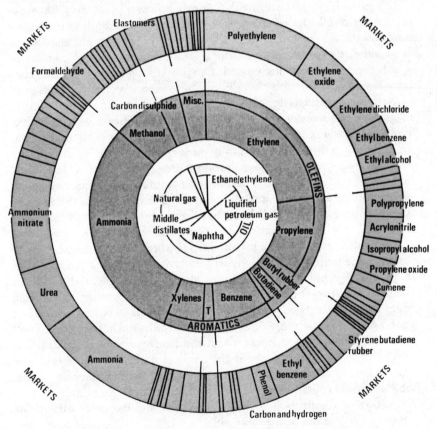

FIGURE 1. Typical petrochemical market in the United States in 1973.

chemical fertilizers promoted high agricultural productivity, which not only increased crop yields but also promoted the production of livestock that convert crops into protein, albeit with marked inefficiency in important cases. (That's why Wiener schnitzels are made out of turkey meat in Israel!)

The postwar era coincided with a massive substitution of natural products by those derived from oil. Natural fibers were replaced in many articles by such materials as nylon, terylene and Orlon. Increasing amounts of synthetic rubber were used to replace the natural product. In wrapping and packaging, plastics replaced forest products such as paper and cardboard, while structural and engineering plastics invaded a field that until then had been the almost exclusive preserve of metal and wood.

Things rarely stand still in the world, however. Not only have the economics of plastics been altered by high oil prices, both at the beginning and during the end-product manufacturing process; but also, at the same time, the producers of natural products are not just content to watch their markets being eroded. One interesting example of this concerns the Malaysian rubber producers. By genetic selection they have improved the productivity of their trees by no less than an order of magnitude. As a result, natural rubber is now significantly cheaper than synthetic rubber. Given that the mechanical properties of natural rubber are superior to those of synthetics in most respects, there is a strong case for using more natural rubber in motor tires. Why is it not done? Is it because of reluctance on the part of tire manufacturers to be dependent on one foreign source of supply—Malaysia? But why should there be one foreign source of supply? If Nigeria, Liberia and, of course, Brazil start now, rubber will be in production in six years in those countries.

Glycerol, of which half a billion dollars worth was produced in 1976, can be made more cheaply by hydrogenolysis of corn than from oil. But there are psychological inhibitions about using potential foods as chemical feedstocks. Nevertheless, with world food production growing faster than world food consumption—even India exported wheat last year—such inhibitions are unlikely to be long-lived. Undoubtedly, I will be asked, "Haven't you heard about the inadequacy of grain distribution? Haven't you heard about the cyclical nature of harvests?" The answer is: I have. But there was no rule about cyclical harvests on the tablets Moses brought down from Mount Sinai; and better distribution is a question of resources (which are available) and the generosity of the haves towards some of the have nots.

There are, of course, a number of factors that limit the rate of substitution of the familiar by the unfamiliar. To begin with, human beings tend to be allergic to change in spite of having accepted some pretty startling ones during the last 25 years. In addition, investment in existing equipment and processes often limits the rate of introduction of an alternative. This applies particularly when there is low economic growth. Also if the increase in the cost of hydrocarbons is only a small part of the cost of the final article, as in the case of many plastic objects, the stimulus to change will be small.

But though the OECD countries may be in a phase of low economic growth, certain less-developed countries such as Brazil possess enormous potential during their high growth rate phase (8.6 percent from 1973–1977). Only 5 percent of Brazil's 3.3 million square miles of land surface (12 times the size of Texas) is cultivated and Brazil's potential for increasing the output of natural products is limited only by a lack of infrastructure, both human and inanimate; a similar deficiency, I believe, catalyzed the tragedy that has occurred in Iran.

An alternative reason for preferring a natural product to one derived from oil concerns the balance of payment. This is why Brazil plans to produce some 20 percent of her motor fuel by fermenting sugar from sugar cane with the production of ethyl alcohol. It will cost about $50 per barrel of oil-equivalent, which, although more than 2½ times the cost of imported oil, is of the same order as the pump price to the consumer.

It may be asked why, since the price of Brazilian sugar is less than half that of the European Economic Community (EEC) beet sugar, Brazil does not export sugar to Europe to earn the foreign exchange needed to pay for crude oil imports. The answer is quite simple: the EEC, which cannot look further ahead than the noses of its French, German and other farmers, does not allow it. This puts a great barrier in the way of Brazil, as well as of other sugar-producing countries.

To continue with the Brazilian example: If it is not being done already, Brazil should study a quite different way of trapping the sun's energy to generate steam and to do other things for which Brazil now depends on oil. I do not refer to the production of ethyl alcohol from sugar cane or cassava, already quite major undertakings in Brazil, but to the conversion of water hyacinth, a floating weed that has no known use, into methane. Water hyacinth has a phenomenal growth rate (up to 87 dry tons per acre per year) and the way to convert it into methane is known. Some economic information about the growth and use of the water hyacinth is shown in TABLE 2 and FIGURE 2. Coupled with the control of the currently uncontrolled Amazon floods,[5] the covering of 0.05 percent of the area of Brazil with water hyacinth would have quite an effect on the

TABLE 2

ECONOMICS OF METHANE FROM WATER HYACINTH
IN AN AREA OF 1,600 SQUARE MILES*

Capital investment	$3.2 billion
Average annual pre-tax profit, 1976–1995	$0.5 billion
Methane production, Standard cubic feet per day	0.9 billion (= 160,000 barrels of oil-equivalent per day)

* Based on Lecuyer and Marten.[4]

economy of the whole of South America. Fifty million barrels of oil-equivalent a year, which that 0.05 percent would produce, is not exactly peanuts. Why not try a few hundred acres and see what goes wrong?*

TABLE 3 lists a number of possible future methods in which agricultural crops could be used to supply products that at present are mainly obtained from oil. Tropical and subtropical parts of the world have an advantage over temperate zones in achieving high yields, and the United States itself obviously has a great potential for such production. The fact that even now Texas can export rice to Asia competitively shows what American agricultural technology can achieve.

Running down the list in TABLE 3, I would like to mention ethylene. Ethylene is easily derived from ethyl alcohol and can be used as a building block, just as it is for traditional petrochemicals. If we consider the superior qualities of wool and cotton, it is hard to believe that synthetic fibers better than the existing varieties could not be developed from natural ones. As a matter of fact it is already possible to confer on fabrics made from natural fibers some of the important characteristics of synthetic fabrics, the so-called "easy care" properties such as "drip dry," crease resistance, and permanent pleat. So will we need to put natural fibers into crackers or use those submicroscopic scissors? Or will superovulation and battery sheep do the trick?

The last item in TABLE 3, wood, is a natural composite with outstanding properties. There is a great deal to be done with wood, not only by improved breeding, cropping and cultivation, but also by preventing, by suitable sealants, the dimensional changes caused by variations in humidity and temperature.

The prospect of developing new products from natural sources does not portend the demise of the traditional petrochemical industry. Crude

* Articles about the conversion of water hyacinth into methane apeared in *O Estado de S. Paulo* on September 13, 1978 and November 1, 1978.

FIGURE 2.

oil and, in the future, coal will for a long time be important ingredients in the creation of high-performance plastics that are both cost-effective and energy-efficient. In response to federal legislation requiring steadily improving vehicle economy, manufacturers are using increasing amounts of plastic in automobile construction. This trend will continue, with an inevitable increase in the demand for petrochemicals as plastics are used more and more for critical structural members. While more energy may go into producing plastic components than into the steel ones that they replace, energy is saved overall if the fuel economies achieved with a lighter vehicle are taken into consideration. Some quantitative idea of what all this means is gained from a so-called "surprise-free" scenario for the year 2020, which was recently prepared by my former colleagues in Shell. They have in mind that 20 million barrels of oil could well be used per day for non-energy purposes in 2020, to be compared with 4.4

TABLE 3

POSSIBLE FUTURE SUPPLY OF PRODUCTS NOW OBTAINED FROM OIL

Synthetic Petroleum Derivative	Possible "Natural" Substitute
Synthetic rubber	Natural rubber from genetically improved stock
Glycerol	Glycerol from hydrogenolysis of corn
Ethyl alcohol	Ethyl alcohol from fermentation of cassava, sugar cane, or ultimately wood
Ethylene	Ethylene from ethyl alcohol
Synthetic fibers	New fibers derived from natural products
Glass-reinforced plastic composites	Improved wood products

million barrels in 1980. I leave you to decide what the effect on these figures will be of the inevitable surprise scenario.

The most important problem—and one to be greatly feared—is not that of finding substitutes for oil when ultimately it becomes too scarce and expensive: it is the inertia of the world's hydrocarbon infrastructure, caused by its colossal size and the apparent impossibility of adjusting it, or even seriously thinking about adjusting it, to accommodate those twenty-first century scenarios. To take an example from the transportation part of the hydrocarbon spectrum, just think what is implied, all over this country, by "filling up" your battery-operated cars at gas stations, let alone what is implied if the energy source, instead of being electricity, heat or "canned fuel," turns out to be hydrogen, which, I imagine, may also be canned. We have probably got 20 or 30 years to adjust that infrastructure—and that is not very long. An inadequate or antiquated infrastructure can undo any country, whether there is an Ayatollah Khomeini or a Shah-en-Shah in charge.

You may have guessed from the title of this paper that I believe bio-energetics and bio-chemistry (and not the biochemistry we learned at school) could make a major contribution to the solution of our future crude oil headaches. But it depends how that contribution is formulated and developed. Let me illustrate that dependence by my own mistakes. Being confident that a trip "back to the bio-board" was right, I was determined, as Shell's Research Director, that that company should get into the race and that our research function should be one of the first to get its hands dirty in this field. But, I thought, "they" (the board) would more easily swallow Rothschild's strange and quite expensive pill if I could justify it commercially from the beginning. In other words, I started some applied rather than some pure research, although the distinction is not a felicitous one. But I underestimated the value of your soybeans in comparison with that of my bacteria, which were making protein out of North Sea methane, and it became clear that my protein was commercially unacceptable. I hope my former employees will not suffer as a result of my short-sighted and pessimistic view about the reactions of their board to my "unapplicable" crystal-gazing. The fermenters are idle now, but a skeleton staff for biocatalytic research remains, for which we are thankful. That, probably, is how we should have started in the first place.

This experience may be relevant to the problems of other research administrators, public and private. For the implication of my views is that we must be very conscious of the need for biological input, both across the board and on the drawing board; that we need to support biological technologies now, not all in good time in the future; and that we also need to devise ways in which some of our current research effort will contribute to the solution of our future economic and social problems. That is not a plea to turn quarks into ploughshares and taus[6] into pruning hooks, nothing as dramatic, unrealistic and unnecessary as that.

I just said that I engaged in crystal-gazing—it is an activity that I find irresistible. But I do not forget what the physicist and science fiction writer A.C. Clarke said: "When an elderly scientist tells you that something is impossible, he is almost certainly wrong."

Acknowledgments

I am greatly indebted to the Royal Dutch-Shell Group of Companies and, in particular, to Mr. E. V. Newlands and Mr. K. R. Williams for major help in the preparation of this paper.

References

1. ROTHSCHILD, LORD. 1956 Fertilization. Methuen & Co. London.
2. BRETHERTON F. P. & ROTHSCHILD, LORD. 1961. Proc. Roy. Soc., Ser. B **153**: 490–502.
3. REYNOLDS, A. J. & ROTHSCHILD, LORD. 1963. Proc. Roy. Soc., Ser. B **157**: 461–472.
4. LECUYER, R. P. & J. H. MARTEN. 1976. Clean fuels from biomass, sewage, urban refuse and agricultural waste. Inst. Gas. Technol.:267–286.
5. PANERO, R. 1972. The Amazon—a catalytic approach to development. HI-1664-RR. Hudson Institute, New York.
6. "Leptons—what are they?" New Scientist **81**:564–66.
7. CLARKE, A. C. 1970. Epilogue. First on the Moon. By G. Farmer & D. J. Hamblin. Michael Joseph. London.

Comments Following Lord
Rothschild's Paper

GERARD PIEL

I WANT TO EMPHASIZE Lord Rothschild's points about taking only the short term into account. With Rothschild, I share an abhorrence for what he calls the taxonomic distinction between basic research and applied research. Consider those reports from the National Science Foundation that estimate the volume of basic research being done in this country. These estimations are based upon some divination of the purity of motive for which the work is done. Perhaps, this is accomplished with a dowsing rod, for there is no visible evidence of the distinctions between *pure* and *applied* that are always made to reassure the public of the value of their investment in fundamental research. I make a simple-minded distinction that does rough justice, I think: I ask how much public money and how much private money is going to the support of science in our universities.

When we look at the record, we see that now, for more than a decade, the constant dollar figure has been down and has only just come level. Meanwhile, the real costs of research in our universities have increased ahead of the inflation rate; the cumulative increase over the past decade comes to between 33⅓ and 50 percent. Effectively, therefore, support of science in the United States has been cut by a third or by half.

This is a matter of the utmost gravity. Of even greater concern, however, are the terms on which public funding has been flowing to the universities. University science cannot be entrusted to support by project grants, especially project grants that flow in greatest volume from mission-oriented agencies.

Fish cannot tell you much about water, and university scientists cannot really tell you much about the inductive forces that have been diverting and distorting what would have been the pattern of a freely motivated university science. What Lord Rothschild had to say about the importance of biology in the future of our energy economy points out one significant distortion. The life sciences in our country have been supported for the past 30 years by the health agencies, with just the tiniest

---

Gerard Piel is President and Publisher of Scientific American.

trickle of funds coming from the Department of Agriculture. In consequence, our country's biological community is preoccupied with mammalian, if not human, biology, and our enterprise in the botanical sciences is all too weak and feeble. The plant world now stands open to the powerful tools that have been developed in the field of molecular biology, but we do not have an adequate proportion of the rising generation trained in or interested in the plant sciences.

In view of this situation, ever since 1965, when the project grant system was at the height of its celebration, I have been urging the universities to enter into open politics with the American electorate and to assert the proper claims of higher education and science on the Federal budget.

It is essential for the health of these institutions that they receive institutional support from which they can deploy venture capital funds, especially to their younger scientists. It seems not only redundant, but offensive as well that when a junior faculty member receives an appointment at one of our great universities—that three- or five-year appointment that gives him or her the last shot at a tenured position in a first-rate university—that young person has got to take his hunting license to Washington and get his work validated by some peer-review panel . At the very least, universities should be capacitated to put new talent in the field that is chosen by their own standards of excellence.

Therefore, I hope that there will be increasing concern to revive the flow of Federal funds to university science and to change the terms on which these funds are given.

United States Science
and Technology Policy

FRANK PRESS

Tʜɪs ɪs ᴀ ᴛɪᴍᴇ when the national policy of the United States with regard to science and technology is a critical factor in many areas of national concern, and when it plays a definitive role in setting the future directions of our country and in our adjustment to new pressures, both domestic and foreign.

On March 27, 1979, President Carter delivered an unusual message on the state of science and technology; this message forms the basis for my comments. It is unusual because it spells out in a single, coherent review and analysis the many aspects of our daily lives touched by science and technology, the decisions we must make in which science and technology play key roles, and the general health and status of our scientific and technological structure.

I will outline briefly the principal considerations that shape our major current activities in the field of science and technology policy. Since the President's message is so timely and so specifically focused on these issues, I will offer several quotations from that message.

First there is the question of the Federal government's role in supporting research and development (R&D). The government plays a part in this in three different ways. In the first place, it supports R&D that directly serves to meet technical needs, for example, those relating to defense and space. Next, it supports R&D where, for reasons of high risk or economically unattractive long-term pay-off, the marketplace will not be effective in developing new technologies to meet specific national needs, as, for example, in the field of energy. Last, it supports R&D—but primarily research—where there are broad economic and social needs that science and technology can help meet.

Basic research has been of particular interest to the Carter Administration. It is not only the quest for an increased understanding of ourselves and the universe, but is also underlies most of our advances today in new inventions, in agriculture, in health care, and in technologies for defense, space, energy and environmental protection. During the late 1960s and

---

Frank Press is Science and Technology Policy Advisor to the President

0077-8923/79/0334-0038$01.75/2 © 1979, NYAS

early 1970s, the United States suffered a serious decline in governmental support of basic research. That decline has now been reversed. And it has been a hallmark of the Carter Administration that there has been a 26 percent increase in basic research in the last two budgets, even in this time of other budgetary stringencies.

Besides working with this increased budget, we are also pursuing many nonbudgetary measures to enhance the conduct of research at universities and provide more opportunity for young scientists and engineers.

We have also been systematically reviewing the programs of the mission agencies of the Federal government to see where and how their missions could be better accomplished through more attention to, and support of, basic research. This effort has produced important results already in the Departments of Energy and Defense.

Another research-related matter that is getting the attention of the Administration involves the dissemination of research results and the transfer of technological know-how. We cannot fully capitalize on our investment in new knowledge unless we move that knowledge out of the laboratories and off the drawing boards into the community, where it can serve industries, the economy and the population. Improving this flow of information and its utilization is a complex matter requiring attention from many perspectives.

For instance, it is important to strengthen the linkage between universities and industry. There have long been beneficial exchanges between them, but we must bolster these linkages in order to stimulate industry with new knowledge from the frontiers of research, and to give universities the benefits that could come from working more closely with industry.

We are furthering another important linkage in our intergovernmental efforts by improving the flow of information and technology between the Federal government and State and local government, and so learning more about the problems that science and technology might help solve through working with governors and mayors to meet the needs of people in those communities.

Still another linkage receiving much attention is that between the Federal government and industry. American industry is currently under tremendous pressure. It faces strong and growing competition from abroad. It faces the need here at home to improve its productivity, to meet the problems of rising material and plant costs, and to deal with new environmental, health and safety demands. A principal way to meet those challenges is through technological innovation, which during the past three decades has accounted for some 30 to 40 percent of the nation's economic growth.

As the President states in his science and technology message:

> Innovative industries are our most productive, create more new jobs, and are the most competitive in world markets. When too few new industries are established, or older ones do not develop enough new products and more efficient operations, the stagnation is reflected in our economy. A lag in productivity worsens inflation. Innovation is essential to our battle against inflation.
>
> More and more countries are industrializing, building industries in which this country once was preeminent. These are countries whose competition is healthy. We welcome their prosperity. We do not seek to limit their growth through tariffs and other trade barriers. Rather, we must seek to improve our own performance through renewed innovation in fields where we excel—such as agriculture, drugs, microelectronics, computers, aircraft, space satellite systems and many other technologies. We also need to make our lower-technology industries more competitive through innovation.

While Americans have not lost their ability to innovate, we recognize that if there is to be a revitalization of industrial innovation in our country the Federal government will have to play a major role. A full range of Federal activities influence and impinge upon industry. Among these are Federal support of R&D, government procurement, patent policy, trade and tariff laws, environmental, health and safety regulation, tax laws, and laws governing the structure of industries. Recognizing the effect of all of these on industry and on its ability to innovate and grow, the President last year commissioned a Cabinet-level study on this subject. Its output is expected shortly. He has promised to give the recommendations the attention they deserve and, together with the Congress, to take action.

In the meantime, the Administration has been moving ahead with a complete range of activities designed to improve regulation and the regulatory and economic climate in which our industries operate. These include the establishment of a regulatory council and the publication of the first regulatory calendar (these will give the public a forecast of upcoming regulatory activity), and a series of innovative concepts in regulation. These efforts will continue, and they should greatly help both to relieve industry of unnecessary regulatory burdens and to achieve a better balance between economic and environmental needs.

Let me next consider some of the specific questions upon which our science and technology policy is focusing, in particular, the increasingly urgent energy problem.

This problem involves closely related objectives:

(1) reducing dependence on foreign oil and minimizing the effects of supply disruptions, with conservation a key element;

(2) implementing programs and policies that encourage domestic energy production and efficient use, without serious inflationary impact;

(3) developing inexhaustible energy sources for sustained economic growth through the next century;

(4) making the transition from primary reliance on depletable oil and gas to predominant use of more abundant energy sources;

(5) developing safe nuclear power systems, which, while limiting the potential for international proliferation of nuclear weapons, will increase our energy supply; and

(6) using all energy sources in ways that do not endanger the environment or the health and safety of our citizens.

If we as a nation are going to advance economically, we must have an assured supply of energy and other essential resources. There *is* an energy crisis. And the sooner we accept this fact and move ahead cooperatively and vigorously, the better chance we will have to head off its worst implications. The greater part of the strategy and program to accomplish this is already in place. It includes providing for increased oil and gas exploration. It includes programs for conservation, and for moving ahead with near and mid-term energy alternatives—for example, the varied uses of coal, enhanced oil production, and solar heating. It includes the future development of an abundant supply of coal and oil shale. And it includes long-term options, such as exploiting the variety of solar technologies and fusion.

While energy has taken the spotlight, we have also focused on other resources. A major Administration study on non-fuel minerals is under way, concerned with both the availability of such resources and ways of tapping new and lower grades of ore with minimum environmental impact.

We are also concerned with agricultural resources. Our agricultural science and technology, combined with good land and a temperate climate, have made us preeminent in world food production. But to maintain that leadership, and more importantly to meet growing domestic and global needs, we must make significant advances in agriculture. These will depend heavily on research in basic crop and animal sciences, on the ability to develop higher yielding crops that have improved nutritional quality, are resistant to pests and weather changes, and are able to grow well with less fertilizer and less energy-intensive cultivation. There are also possibilities for developing crops that will turn arid lands into economically productive areas. In addition, we are working in agricultural research directed toward the needs of developing regions, both arid lands and the wet tropics. And we are considering ways to bring new science to the ancient technique of aquaculture.

Food and nutrition are of major importance in themselves, but they represent only one aspect of the broader concern for human health itself, which is a great concern of this Administration, not only because of all that good health represents to individuals and to the country's well-being, but also because of the rising cost of health care. This cost has in many respects overshadowed some of the remarkable advances made in dealing with many diseases. But perhaps more significant is the fact that it is shifting our attitudes toward, and emphasis on, disease prevention, a need that calls for better biomedical research to gain knowledge of both the basic causes of diseases and of ways to counter them in the earliest stages, if not to eliminate them altogether. The preventive approach, typified most dramaticaly by development of a polio vaccine rather than relying on iron lungs, could save billions of dollars in health care costs, and alleviate much human misery as well.

In carrying out such a health strategy, new work is being supported in molecular biology, the neurosciences, behavioral sciences and genetics. We are emphasizing research in reproductive biology and the underlying mechanisms of normal development and disease. Research and service are expanding in such currently troublesome areas as unwanted pregnancy, smoking, and alcohol and drug abuse among adolescents. And we are placing more emphasis on the causes of common disabling conditions such as diabetes, arthritis, and neurological and digestive diseases. Finally, increased attention is focused on mental health.

My office is very active in these concerns, and works closely with the Department of Health, Education and Welfare, the Department of Agriculture, the National Science Foundation and other agencies on health and biomedical problems. In addition, we are involved with the international medical community in matters of world health, which is of great interest to the President.

Another important area of science and technology policy is space policy. Two decades ago we entered the space age. Ten years ago this July we landed on the Moon. We have also explored the surface of Mars, probed Venus, and closely observed the moons of Jupiter. These are remarkable accomplishments. The question is: Where do we go from here?

The Administration is responding with a comprehensive space policy. It is a policy that will enable us to begin to reap major benefits for our people and for people everywhere from the more than 100 billion dollars that we have already invested in space. In addition, it will continue the exploration of our solar system and the universe, furthering that great scientific and intellectual quest.

In the new space policy, earth-orbiting satellites will play major roles in communications, weather forecasting, navigation, the monitoring of agricultural resources and the environment, the discovery of new mineral resources, and a host of other activities related to an improved understanding and management of land and water resources. Satellites will also figure prominently in national security.

With the advent of the Space Shuttle we are entering a new era. The Shuttle will vastly improve the economics of space activity. It will serve as a national space transportation system for decades to come. It will provide flexibility, reduce costs, improve national security, and make possible a new level of international cooperation in space. We also foresee a role for it in making possible the beginning of space industrialization. When the Shuttle is operational we may be able to lay the groundwork—or "space work"—for some of the more ambitious, more visionary projects that are being conceived.

The view from space, of course, adds to knowledge of the forces continually at work on this planet. Lately we have not had to be reminded about such forces. This has been a year of devastating effects from such forces in the form of floods, earthquakes, tornadoes, volcanic eruptions. All these have reinforced the need to acquire a better understanding of the causes of these destructive forces and of better ways to reduce their devastating impact on people's lives.

Scientific advances in geophysics, meteorology, and climatology have improved understanding of natural phenomena, and will continue to do so. But our predictive capability in these fields is still limited and needs improvement. We also need to improve engineering skills, planning capabilities, early warning and communication systems, and other economic and social activities related to mitigating the hazards and effects of natural disasters.

The Administration is committed to policies and programs to accomplish such aims. We have been working to implement the National Earthquake Hazards Reduction Program. This effort involves Federal agencies, experts in universities and the private sector, and state and local governments in improving our understanding of earthquakes and our reactions to earthquake warnings.

The President has recommended, and the Congress has approved, the creation of a Federal Emergency Management Agency to coordinate Federal programs which assist areas and individuals affected by disaster emergencies.

We have also stepped up research on climate, with a major effort or-

ganized under the National Climate Research Program. Much research is
being conducted in connection with the problems of atmospheric ozone
and carbon dioxide, and the alteration of oceanic flow patterns.

Let me turn next to international affairs. As the President has stated:

> Science and technology is increasingly international in its scope and significance.
> This international dimension affects the planning and conduct of our research and
> development activities. Such activities, whether carried out by us or by others, serve
> to increase the fundamental stock of human knowledge. They can also foster com-
> mercial relationships, impact on the quality of life in all countries, and affect the
> global environment. Both our domestic planning and our foreign policy must reflect
> an understanding of this wide-ranging impact of science and technology.

The Administration's policy in international science and technology is
shaped by several themes. We're pursuing international initiatives that
advance our own research and development objectives. We're strengthening exchanges in science and technology that bridge political,
ideological and cultural divisions between countries. We're formulating
programs and projecting plans for institutions that will help developing
countries to use science and technology. And we're cooperating with
other nations to manage technologies that have global impact.

Let me mention some examples of these activities.

As the cost of large-scale research programs and facilities rises, the
financial support of these becomes more burdensome. It is therefore advantageous to join with other countries in cooperative research programs. We have participated in such programs before, and are doing so
increasingly today. We are continuing in the Global Atmospheric Research Program and the Deep Sea Drilling Program. We are discussing,
with other nations, an extension of the latter to a new program with a
new ship, the Glomar Explorer, to carry out drilling into the margins between the outer continental shelf and the ocean basin. With the European
Space Organization we are planning a mission to examine the polar
regions of the sun.

In the important field of energy we are cooperating on research and
development projects with the International Energy Agency, and have
explored the possibility of joint efforts with the European Economic
Community. We have also reached an agreement with Japan to conduct
cooperative research in nuclear fusion and synthetic fuel production
from coal.

During the President's visit to Mexico, an agreement was reached on
intensifying scientific and technological cooperation on matters of
mutual concern. One area ripe for such cooperation is research on new
crops suitable for the arid lands between our countries. Research and
development in the field of water resources is another, as is the problem

of housing and urban planning in the border region between the United States and Mexico.

A significant event during the past year has been the normalization of relationships with the People's Republic of China. Closely involved with this is China's desire to modernize with the help of scientific and technological exchange with the West. We are cooperating in such an exchange in many areas of concern: agriculture, energy, space communications, high-energy physics, health, and others, and we are taking part in an exchange of students and scholars as well. All of this provides a sound beginning for increased technical, economic, and social ties between these countries.

As the President indicated in his message to the Congress:

> Our scientific exchanges with the Soviet Union are of special significance. At the sixth meeting of the United States-Soviet Joint Commission on Science and Technology in Moscow in February 1979, we agreed to add new cooperative areas of interest to both sides. I expect to see continuing improvement in the quality of our exchanges with the Soviet Union. I also expect these programs to support and remain compatible with our overall political relationship.

A major goal of the Carter Administration in international affairs is concerned with helping developing countries to help themselves through science and technology. This is a new emphasis in the field of development assistance. It is one that must succeed if the more than 800 million people in the world who live in dire poverty—and whose number is growing—are ever to raise themselves above the level of bare subsistence. Only by helping these people to become more productive will development really take hold and enable the world to achieve the economic opportunity essential to world stability and peace in the long run.

Recognizing this important need, the President has sought the creation of an Institute for Scientific and Technological cooperation. Such an institute would be devoted to working with developing countries in an effort to improve their scientific and technological capability. It would work with the Agency for International Development (AID) and other government organizations, and with universities, private foundations, and industry to meet the needs of development. It would work with the "middle tier" countries as well as the poorer ones.

Much more can be said about international cooperation, but I will conclude with a few words about national security, a subject very much on peoples' minds as we approach the final stages of activity with regard to SALT II. National security is largely a matter of high technology these days, whether one is talking about advanced weapons systems or the capability to verify adherence to an arms limitation agreement. As the

President pointed out in his message to the Congress, we must use our science and technology is ways that ensure that we are: (1) maintaining technological leadership in weapons systems; (2) utilizing technology to reduce defense costs; (3) building a defense research base for future national security; (4) preventing exports of products and processes that would erode our security; and (5) using advanced technological capability in the pursuit of arms limitation.

These are difficult and complex responsibilities, but all must be pursued. We are doing so in the belief not only that our military strength should be second to none, but also that every reasonable means should be sought to slow down, and eventually halt, a nuclear arms race. This second belief emphasizes the importance of SALT II.

In reviewing some of our science and technology policy and programs, I know I have not covered every aspect of our involvement. I do hope, however, that you have some feeling for the scope of work undertaken and for the way in which the President and the Administration regard the role of science and technology in our society. Much of the country's future depends on the positive and constructive way that political, industrial, and civic leaders feel about and act toward these two great forces in our society. Although there have been some setbacks and exceptions, I am encouraged by the ways most of these leader view science and technology and support them. I think that this bodes well for our future.

Technical Advance and Economic Growth: Present Problems and Policy Issues

RICHARD R. NELSON

THE OECD THEN AND NOW: WHAT IS DIFFERENT?

To SKETCH, from the perspective of somewhat heterodox
economist, some of the key issues of science and technology policy that
face the countries of the Organization for Economic Cooperation and
Development and particularly the United States, and that call for new
thinking over the coming decades, my point of departure will be to
describe a number of important differences between the OECD countries
as they are now, and as they were during the heyday of rapid growth. To
some extent, these differences are describable in terms of the economic
troubles of the 1970s—slow productivity growth, high inflation rates,
high levels of unemployment, and balance of payments stress—compared with the more salutary economic conditions earlier. But such a re-
counting provides a biased and somewhat superficial view of what has
changed in the OECD countries over the past quarter-century.

Compared with the 1950s, virtually all of the OECD countries now
have much higher real per-capita incomes, and levels of real private consumption have risen dramatically. Virtually all of the OECD countries
have taken major steps to place a floor under individual and family standards of living, and to protect people to a significantly greater extent
than before from the vicissitudes of labor-market conditions and personal infirmities or bad luck. As a striking example, a large fraction of
the currently unemployed in many of the OECD countries have had their
income losses largely buffered by unemployment payments. In addition
to providing income maintenance, the modern welfare states have vastly
increased the range of goods and services that are public provided or

*Richard R. Nelson is a Professor at the Institution for Social and Policy Studies
and Professor of Economics at Yale University.*

0077-8923/79/0334-0047$01.75/2 © 1979, NYAS

which are largely financed through government rather than private budgets. Almost all of the OECD countries now have some form of national health insurance. Most provide housing and other services at subsidized rates for individuals and families on low incomes.

As a consequence jointly of generally higher average living standards, and of the greater role of governments in influencing income distribution and resource allocation, the old traditional sectors of agriculture, mining, and traditional public utilities now account for a significantly smaller fraction of the work force than they did 25 years ago. Private and public services, in particular, now absorb a significantly larger fraction of the work force.

The growing governmental role in resource allocation and income distribution has been accompanied by increased regulation of private economic activity. Environmental and energy regulation, and job safety and product-attribute regulation, are now forces that business must heed in deciding what they are to do. These forces constrain and complicate the answer to the question of what can be produced and sold at a profit.

I propose that the various developments sketched above are linked together. They reflect growing affluence and an associated rise in the relative importance of values that the rapidly growing, basically market-guided economies of the 1950s and early 1960s were slighting or punishing. I will argue shortly that the described developments have something to do with the observed economic malaise of the past half-dozen years. But they need to be recognized in their own right. Even if the economic sailing had remained smooth, the developments would pose challenges to science and technology policies. Given the rough sailing we have experienced, the challenges are compounded.

The Economic Slowdown

The last half dozen years look particularly bad in comparison with the era of rapid growth that marked the previous decades. But to put that perception in context, it is important to recognize that economic progress during the 1950s and 1960s was unusually rapid by historical standards. In order to understand what has gone wrong recently, it is useful to consider what went right earlier.

Certainly the initial conditions set the stage for rapid growth. Almost all of the OECD countries had experienced economic stagnation during the thirties. Many of them experienced devastating physical losses during World War II. Thus, all of the nations came out of the war with a perception of large unmet needs for private and public goods, of capital shortages, and of a large shelf of unused technical advances that had accrued over the years of depression and war. This was a situation virtually guaranteeing a high rate of return on investment and rapid productivity growth as capital expanded, if aggregate demand and supply could be kept in relative balance.

In view of contemporary tendencies to depreciate economics as a science and economists as policy advisers, I would remark that, throughout the post-World War II period, economists have advocated, and governments have gone along with, active demand-management policies and, until recently, these policies have worked. Aggregate demand growth was not permitted to fall behind growth of the capacity of economics to produce goods and services. Until recently, unemployment was generally kept low, and inflationary forces generally under control. It well may be, of course, that the initial conditions made the balancing task of fiscal and monetary policy relatively easy. Because of high investment demand, governments did not have to act directly as demand stimulators and did not have to run budget deficits to that objective. High investment and rapid productivity growth meant that pressures in the high-employment economy toward sharply rising wages did not have to result in high inflation rates. In any case, until recently, macroeconomic policies worked.

The era of rapid productivity growth was marked not only by high investment rates, but by significant changes in the allocation of labor across sectors and industries. While high levels of employment were preserved generally throughout the period, in some industries employment fell rapidly. But with overall demand high and growing, and overall unemployment low, labor mobility was facilitated. In the first part of the rapid-growth era, the dominant shift was largely out of agriculture into manufacturing, transport, public utilities, and services. Toward the end of the era, the dominant shift was into services.

While all of the OECD countries entered the post-war era with a catalogue of technologies that had been developed during the 1930s and the war years and that had not been incorporated into practice, this source of productivity growth with high returns to capital sooner or later had to run out. But research and development spending in all of the OECD countries was strikingly larger during the rapid-growth era than in earlier periods. If the initial conditions set the stage for rapid growth and made the task of demand-management policies relatively easy, the significant technical advance experienced during the postwar period continued these salutary conditions. The rate of return on investment continued to be high (although late in the 1960s there were some signs of decline). Investment rates continued to be high, and labor continued to move from low to high productivity sectors. By the late 1960s, the bulk of employment growth was no longer in manufacturing but in the services.

This shift into the services can be explained by two related but different factors. First, in the late 1960s, private demands for both manufacturing and services seem to have been relatively price-inelastic. While sectors experiencing rapid productivity growth did experience relative price declines, since demands were relatively insensitive to price, resources shifted out of industries with rapid productivity growth, principally manufacturing, and moved into those experiencing slower productivity growth, in the case of services. The second factor was a significant increase in governmental spending on services, particularly education and health, but including as well a wide range of welfare services.

By the late 1960s, there were some indications of trouble on the horizon. The United States began experiencing both inflation and balance-of-payments problems, and there is evidence of deceleration of productivity growth at about that time. One prominent study of that early deceleration in productivity growth accounts for much of it in terms of a shifting allocation of resources. There was little evidence of any sector-by-sector or industry-by-industry decrease in rates of productivity growth. The inflationary problems of the United States were contagious and some of the basic problems that were leading to inflation in the United States obtained as well in Europe. By the early 1970s, the European economies, too, were showing problems in restraining inflation. However, prior to 1973, the productivity-growth deceleration that marked the United States had not set in widely in Europe.

1973 clearly marks a break in trend. Since 1973, productivity growth has been slower in all of the OECD countries. Further, it has been slower sector by sector as well as in the aggregate. And, slow productivity growth has been accompanied by much more intransigent inflationary pressures and significantly higher unemployment rates than obtained earlier.

What lies behind this unhappy syndrome and why is it proving so stubborn? Economists certainly are not in full accord on diagnosis. I would urge consideration that the dominant proximate cause has been governmental policies, virtually universal throughout the OECD countries, that have been trying to restrain inflation by restraining growth of demand. These policies have been relatively successful in restraining demand growth. But, as a consequence, supply growth has been restrained as well. Until recently at least, there has been limited pressure on capacity and hence limited incentives for new investment, together with stagnation in employment. Particulary in the European countries, labor legislation has protected jobs. Productivity growth has always tended to decline when economies slid into recession. Job-protecting policies have increased this tendency.

I don't mean to say here that the root of the problem is pernicious government policies. Government policies are what they are because inflation has been so rapid, and because there is widespread political pressure on governments to do something about inflation. I do mean to say, however, that these policies, whatever their effect on restraining inflation, also have restrained productivity growth.

Let me push the analysis one stage further. Why the inflationary pressures now and not earlier? Analysts have cited a long list of factors. There are several that I think particularly germane to the present discussion.

First, I would argue that the rapid rise of living standards during the 1950s and 1960s sharply raised expectations of what might be achieved in the future, and actually intensified squabbling about how resources were to be allocated and incomes distributed. While the dispute about priorities during the Vietnam War made the allocation struggle particularly visible in the United States, throughout the OECD countries questions about whether increases in gross national product (GNP) should be allocated to public services, or to the provision of private goods, were the stuff of politics. Throughout the OECD countries there seems to have been widespread verbal agreement that the real incomes of lower-income individuals and families should be brought closer to the mean. However, when push came to shove, middle- and upper-middle-income individuals did not put up with a reduction in their relative shares. The dispute was fought out both in the arena of private wage-bargaining, and in the arena of welfare-state programs. And, by the late 1960s in the United States, and somewhat later in Western Europe, the political pressure for environmental and safety regulation, which had been building up for some time, reached sufficient force so that these demands and requirements also were placed upon the economic system.

I do not want to take even a hint of a stand here regarding which of the demands were (and are) socially most important. I only want to propose that by the late 1960s, the increasingly politicized economic systems of the West were generating demands faster than could be met even with the rapid growth in productivity being experienced then. And when, as a result partly of restrictive policies and then the great shock of the oil embargo and the oil price hikes, real output growth in the OECD countries slowed down, the problem was compounded. I do believe that to get their economies working reasonably again, the OECD countries are going to have to be more effective than they have been in balancing competing claims against available means.

But even were we to achieve this, there are two salient structural differences between the present and earlier economies that will make rapid productivity growth difficult to achieve again. I referred to both of them earlier. We have been experiencing a shift in the allocation of resources away from manufacturing and public utilities to private and public services. And through regulation we increasingly are forcing attention to non-market values in market sectors.

Rates of productivity growth in the service sectors historically have been very slow. This is partly a measurement problem, but almost certainly real rates of productivity growth in services in general have been less than in manufacturing. Thus, this shift in the allocation of resources brings down the average rates of productivity growth even if all the sectoral rates are maintained.

Regulations force attention to values that otherwise would not be recognized. Analysts differ on the costs associated with these new regulations. However, they surely increase resource costs in traditional dimensions, and shift the allocation of investment in new plant and equipment, and in research and development (R&D) toward meeting these new values and away from enhancing productivity as traditionally defined. This isn't to say anything negative about these regulations. These shifts in allocation are after all, their intent. But as we make these values count more, we well might expect slower expansion in production of goods and services as traditionally defined.

Both of these structural developments pose significant issues for science and technology policy. So does the continued economic stagnation, which I propose is acting as a deterrent to traditional industrial research and development.

Note that I am not proposing here that an earlier slowdown in technical advance was an important source of the current economic malaise. Some other analysts have proposed this. It has been remarked that R&D spending in the United States began to trail off in the late 1960s, well before the 1973–74 crunch. But the national R&D cutback is completely accounted for by declines in defense and space R&D. Most econometric studies of the sources of productivity growth have been unable to attribute much weight to spending in the areas of defense and space R&D. And private industrial R&D expenditures, which earlier econometric work has shown to be strikingly related to rates of productivity growth, held up quite well until the mid-1970s.

Since 1973, there has been significant deceleration in R&D expenditures, private as well as public. But for the most part, this is associated with deceleration in growth of real output, rather than any reduction in the ratio of real R&D expenditure to real output. There has been recent evidence, however, of a shrinking of time horizons and a growing conservatism regarding industrial research and development. This development, like the slowdown in R&D spending, is just what one would expect given the slack economic conditions.

The slowdown and growing conservatism of R&D spending, along with the shift in allocation of resources toward services and the new regulatory regimes, all pose serious problems for R&D policy. Let me now explore these.

POLICY ISSUES REGARDING TECHNICAL CHANGE

In the preceding section I argued that the current stringency in R&D budgets, and the growing conservatism of R&D, stem from the current economic malaise. They are a consequence rather than an initiating cause. But slow and conservative technical advance can make it harder to break out of the current rut. And faster and more innovative technical advance may make it easier to get out.

More rapid technical advance can facilitate more rapid productivity growth, and hence enable wage-increase demands to be met with less inflationary pressures. I would propose that more rapid technical advance and productivity growth, far from being a threat to employment, would facilitate a reduction in unemployment. I argued above that the proximate source of today's high unemployment is restrictive government policies. If more rapid productivity growth can facilitate better control of inflationary pressures, governments will be able to relax their restraints. Thus, more rapid productivity growth may be a prerequisite for a return to higher levels of employment.

Many scholars have argued that support of long range research and exploratory development in the basic technologies that underlie a wide variety of industries, is a feasible and potentially fruitful role for government. In a few fields, such as atomic energy and aviation, the American government has pushed such policies. European governments have developed them across a broader front. Experience with these programs suggests that there is a danger of governments' getting too closely involved in making decisions as to which particular products are to be developed or which technologies are to be brought into practice, rather than concentrating on broadening the range of alternatives seriously being explored. But if these proclivities can be kept in mind and guarded against politically, then, assuming such government programs were a good idea during the 1960s, they will be an even better idea during the 1980s. The current slump and the adjustments made in industrial R&D reveal just how fragile private support of exploratory and long-range work is. It would seem an excellent idea if governments took an explicit responsibility for overseeing the adequacy and diversity of the basic technological efforts as they long have taken a responsibility for overview of basic academic scientific research.

I do not want to push the idea that aggressive government stimulation of basic technological work can be a central instrument in resolving today's macroeconomic problems. But I think it should be more widely recognized among economists that further erosion of basic technological progressivity can make today's problems worse, and that policies to stimulate greater innovativeness can be important parts of a salutary package. Over the long run, protecting basic technological effort from the ups and downs of general economic activity may be an important component of a refurbished economic stabilization policy.

My other remarks about policies relating to technical change relate to more microcosmic problems. Throughout this essay I have pointed to two major structural changes in the OECD economies—a significant increase in the fraction of resources allocated to private and public services, and a great enlargement in the scope and strength of regulatory regimes. Both of these developments pose important questions regarding policies relating to technical advance.

The evidence is clear enough that the new regulatory regimes are having a significant influence on technical advance in the sectors most affected. This was their purpose. But it also is apparent that there are significant problems involved in trying to redirect private R&D expenditures through the kinds of regulatory instruments we have been using. In the first place, uncertainty regarding future regulatory requirements may deter firms from trying to develop new products and processes significantly different from present ones, since the regulators may respond to their advent by prohibiting them. There is some evidence that this has been happening. Second, the current kinds of regulations prescribe certain dimensions of environmental insult on work hazard, but not others, and establish particular required levels of achievement with no reward for surpassing them.

Current regulatory standards tend to be set with some notions regarding the costs of meeting them and the responses they will evoke (which may or may not be justified). Sometimes the result is that the standards are met at high costs with very few social benefits; sometimes the result is that the standards are challenged and the regulators forced to back down.

At the least, cases like that of automobile-emissions control indicate that a rather strong governmentally-funded R&D program is necessary simply to enable standards to be set sensibly. Government-undertaken or funded R&D also is necessary if government agencies are to avoid being outclassed by the firms they are regulating in discussions regarding technological options, likely costs, and reasonable expectations regarding the performance of the companies in question. I would tentatively suggest that public responsibility for the funding of R&D aimed at furthering non-market values might go considerably beyond this minimal role of obtaining information. Regulations might well be set more sensibly if Congress and government agencies had to face some of the R&D costs of meeting them.

Another challenge for policy regarding technical advance is posed by the shift in the allocation of resources away from manufacturing industries to other, principally service industries, where measured productivity growth had in general been slower. While in some of the service sectors technology has changed very slowly if at all, slow productivity growth in the service sectors now exerts a bigger drag on income growth than it used to. Significantly enhancing productivity growth in such sectors as construction, urban mass transport, health maintenance, and education, is now even more important than it was before in order to achieve social gains in these areas.

Simply shifting resources into these sectors will tend to draw greater R&D attention to them. But it is highly unlikely that market forces alone will be able to affect major improvements in the technological progressivity of sectors like these. And it is doubtful that government ought to stand idly by and defer to the market for effective R&D efforts. In the first place, the organization of many of these sectors is mixed, with governments playing a large role on the demand side and often on the supply side as well. Directly or indirectly, governments have a great deal of leverage over R&D in such sectors as public housing, health, education, and urban mass transit. This leverage ought to be exerted self-consciously and intelligently. Second, in many of these sectors, like private housing, demand has been protracted, high, and growing, but not much as happened. Government action has the chance of getting something to happen.

As in the case of regulation, for the sectors where the government is heavily involved as demander or supplier, the public agencies involved should spend at least enough on R&D to know what the technological options are and what they are likely to cost. But as with the regulatory cases, I propose that here too the fruitful public role for R&D should go well beyond that minimal requirement, and should involve government finance or cost-sharing in a wide range of R&D activity. In fact, R&D spending in these fields has increased significantly in the United States over the past decade. But satisfactory arrangements have yet to be worked out.

In the case both of R&D focused on the new regulatory regimes, and R&D focused on public services, the advisability of an expanded government role is supported both by its natural access to relevant information and by the legitimacy associated with an acknowledged public responsibility. It will be harder for governments to establish an effective, active R&D policy where there is no strong recognized public responsibility for the values or the services in question, particularly if private suppliers view each other as competitors. Perhaps the most successful example of public funding and subsidization of private-sector activity is to be found in agriculture. Another is public support of the basic sciences and of much of medical technology. In both of these cases public involvement could be justified politically by appeal to the role of the state in assuring that basic needs are met. But roughly the same kind of argument also could be, and was, put forth in the various efforts in the 1950s and 1960s to mount public programs in support of housing technology. The proposed policy departures never were actively implemented. The latter case is differentiated from the former ones, it seems to me, by the fact that suppliers of inputs into housing construction considered each other as rivals. Public support of R&D relating to housing therefore was viewed as posing a sharp threat that some firms and industries would be helped and others hurt. This perception undermined the legitimacy of government programs and was sufficiently effective to keep them bottled up in legislature.

As I understand it, the current struggle with respect to the government's role in energy-related R&D reveals a similiar syndrome. While a broad public responsibility is not questioned, the issue of where public responsibility begins and private responsibility ends involves not only questions of who is in a better position to make decisions and where private incentives are strong and weak, but also the constraint that public funds should not significantly upset the balance of private competition. But, the requirement that public research and development support not upset the private balance can come close to a constraint on public R&D not to generate anything significantly different from what private R&D would have come up with in the first place.

I flag this problem because I think it is a serious one. My arguments herein call for a significantly greater role for public decision-making and public funding of applied R&D. While such a role will inevitably benefit most of us, it will erode the interests of some of us. All of us concerned with identifying the important science and technology issue of the 1980s should be very aware that it is one thing to identify and argue rationally for a new set of policies, but it is something else again to get these policies accepted politically without emasculation.

Major Issues Facing the
OECD Countries

BERNARD M. J. DELAPALME

A**T THIS TIME**, and for the foreseeable future, the problems facing the industrialized countries are formidable. To gain a perspective on how we might address these issues, I propose to present the viewpoint of the Group of Experts at The Organization for Economic Cooperation and Development (OECD), that of industry, and in cases that demand more candor and directness than is possible in official circles, that of the author.

The opinions of the OECD Group of Experts were recently summarized by Professor Freeman of Sussex University and are listed as follows:

(1) The context in which we are now called to act is radically different from that of the last thirty years, during which time there was tremendous development of our economies. It is very important to recognize this fundamental change since it will cause us to alter substantially our way of thinking and acting. It may, for instance, lead us to realize that the classical laws of economics are no longer sufficient to guide our behavior and decisions. The best known examples of this are the insufficiencies of the present notions of the gross national product (GNP) and national accounting. This does not mean that these laws are obsolete, but that something new must be added. Among the new factors, one can cite the growing impact of public regulations, the development of the Third World and its ambitions, a new balance between work and the quality of life, the extraordinary development of communications, and the behavior of raw-material producers, notably oil producers. These new elements are generally accepted. But what is not yet realized is that the con-

Bernard M. J. Delapalme is Vice-President, Research and Development of Elf Aquitaine, France. He is a former President of the European Industrial Research Management Association.

0077-8923/79/0334-0058$01.75/2 © 1979, NYAS

sequences of these developments may be substantial, and will in turn demand conspicuous reorganization of our thoughts and habits.

(2) A vital part of this reorganization is the perception of technical problems as central to our situation, not merely peripheral. This concept is essential, not only because science and technology have played a major role in creating our present societies, a point not widely enough recognized, but more importantly, because potential technical advances will provide solutions to both present and future problems.

In an introductory speech to the Conference of the European Industrial Research Mangement Association (EIRMA) entitled "Technology 88," I cited two examples of ways in which technological progress has solved a problem of meeting higher demand through increased production. Between 1950 and 1976, the number of cars in use in Europe grew from 5 million to 100 million, and between 1966 and 1976 the production of color television sets increased from 5 million to 25 million a year. During the same period of time, the price and the size of basic electronic components was reduced by a factor of 100.

These examples cannot be denied. Yet many people declare that our current level of progress is generally inadequate, and complain whenever 1/10th or even 1/100th of the resulting benefits are suppressed, even though current worldwide shortages often necessitate their suppression. Modern science and technology offer a cornucopia of possible solutions for many of our future problems, and this potential contribution must not be underestimated, as was done in the Club of Rome's pessimistic first projection of the future.

Acceptance of the basic idea that science and technology hold a central and universal position in our world has several logical consequences, as will be discussed later.

(3) If we accept the fact that technical change exerts great influence on the present situation, and may have a strong impact on our future due to the considerable changes in our socioeconomic environment, it will be clear that we must alter both the rate of technical change and its direction.

Technical change depends heavily on the amount of funding available for research, but more than total quantity, it depends on the effectiveness of the research organization that allocates the money. The main reason an increase of money is necessary is to reallocate funds between sectors. This would not only boost the morale of many presently disheartened researchers, but would also bring about the desired change of direction of technical research.

Until recently, the efforts of a country or a company in the field of research were measured as a whole, little attention being paid to allot-ment differentials between fundamental and applied research, develop-ment and demonstration, or between military, prestige, and "bread and butter" areas. Moreover, when a distinction was made, there was fre-quently a misconception about the definition of fundamental research, or of high-technology sectors. For instance, electronics and chemicals have been viewed as high-technology areas, while petroleum and housing have not. These distinctions are no longer valid, and to continue to favor them may lead to an inadequate framework for decision-making.

With respect to the problem of change in the direction of technical research, there has been much discussion of "technology push and market pull." After careful consideration, I believe this discussion is un-fruitful since both aspects are necessary, and the solution lies in the choice of a good combination of both: that is, choosing the most effec-tive and fast-growing technologies (in chemicals and electronics, for ex-ample) and fulfilling the most important and rewarding needs (in the fields of health, food, energy, and housing, for instance.)

(4) With regard to the problem of markets, the importance of public policies is increasingly evident in those fields in which the classical laws of the free market cannot play their role. This may be the case in the field of public transportation, or other town infrastructures, or in the dif-ferent, but very important, field of production of goods by and for lesser developed countries. Another example can be found in the field of health care. In his 1979 State of the Union address, President Carter stated: "We must act rapidly to protect American people against the growing cost of health care, which increases at the rate of 1 million dollars an hour, and doubles every 5 years." The apportioning of expenses among prevention (especially nutrition), pharmaceuticals, and medical care at home and in hospitals, is far from optimum. We have the necessary scientific basis for developing those techniques that would allow for a better financial balance in health care, but the laws of the free market would slow down their development significantly. Society can ill afford this delay. Therefore, it is the duty of public policy to provide necessary incentives for accelerating the process.

In focusing so strongly on public policies, the position of the OECD Group of Experts should be kept in mind. The purpose of OECD is to make recommendations to governments. The position could be taken that governments should do nothing—that they should adopt a *laissez-faire* or *laisser-aller* attitude. But this easy way would certainly not, in many cases, be the most effective.

(5) If we accept the idea that public policies are crucial to our develop-

ment, we must also recognize the dilemma faced by governments in choosing effective policies, especially if those choices must be based on political assumptions, as well as on long-term, ill-defined, and risky options.

There are those who believe that the study of history can provide guidance in judging what kind of process is good, what kind is bad. However, the example of atomic-energy development in France, which some consider good and some bad, indicates that historical instances can be ambiguous, and cannot be trusted to provide reliable guidelines by which to choose correct processes in the future. The case of the Concorde is also not clear-cut, because supersonic transport may finally become a good market offering for Europe—one that would not have been available if France and England had not taken risks for its original production.

It is evident, then, that those in charge of establishing public policies have to find new methods, new concepts, and new patterns of behavior in order to be effective in our profoundly different and difficult times.

If it is clear that classical market laws are no longer sufficient to help solve our problems, it is not less evident that classical behavior of governments and administrations is no longer satisfactory to get things done. But then again it is often very difficult to determine what to do. Often when policy advisers to the government become the policy makers, they discover that their own sage advice does not in fact work.

Even if the choosing of policies and actions has no reliable precedent, it is the duty of the OECD Group of Experts to offer some indications of what might be the best way to enhance progress. I would like to discuss ten such indications. Some suggestions will seem commonplace, but perhaps this is an asset, since governments traditionally fear to contradict public opinion, and might in fact be encouraged to act if they are also assured that reasonable men concur with the advice that they are given. Other recommendations are more original, and even if they are not adopted immediately, they can initiate processes that might finally lead to a real and beneficial innovation.

1. *Greater support of fundamental technology.* The term "fundamental technology" implies that although a research project may be basic, it must be designed to contribute to many practical uses. Therefore, fundamental problems must be emphasized for study, such as those relating to fermentation, aging of materials, reliability, surface states, and so forth. Since these problems may seem too "applied" for universities to tackle, and too risky—that is, too long-term or ill-defined—for industry to study, governments must find incentives not only to make these fields

attractive for study in depth, but also to ensure that their investigation be conducted by the best minds available. This last point should be emphatically underscored, because no good comes of pouring money into fundamental research unless persons of the highest recognized expertise do the work.

2. *Sufficiently extensive introduction of small firms into the innovation process.* Small companies often have a dynamism that is missing in larger ones. They may also have marketing imagination which gives them special mobility and inventiveness. But usually they do not have the advantages of big companies such as worldwide network, critical size in research organization, knowledge of a whole range of markets, financial power, relations with a government administration, and so on. It would be productive to find ways in which the innovative characteristics of the large and the small companies could be made mutually beneficial. The United States has experience in implementing such desirable interaction, but France has done less. However, the Elf Aquitaine Company recently took steps in this direction by offering the use of its sophisticated research center to small local companies. The OECD, incidentally, also suffers from profound differences in the size and capabilities of its member states, and could, as a whole, benefit from some sort of the aforementioned interplay on a country-to-country basis.

3. *Innovation in non-market sectors: the importance of social science.* Productivity is increasing more rapidly in industrial sectors than in the social-service sectors. However, it is probable that good use of modern electronic systems could enhance social-service productivity.

Progress might be achieved more readily if industry, government, and those responsible for the administration of programs made wider use of techniques available in the social sciences. The term "research" as it is generally accepted now by chief directors of research applies only to the fields of the exact sciences, and excludes almost completely those of the social sciences. But both disciplines are essential for innovation. Someone mentioned to me that it was far more difficult to build new towns than to put a man on the moon. Unfortunately, it is just as important for our future to build new and better towns, and it is our responsibility to find ways to do it. This is likely to involve a good understanding of the relationship between the social sciences and the exact sciences:

4. *Innovations in market sectors.* It is just as important to favor innovation in the free-market/private sectors as it is in the non-market sectors. Recently, an interesting initiative was taken by the business-growth service of the General Electric Corporation, along with a number of other companies, to acquire new technologies (products and processes)

through a licensing system. Another significant example is the recent creation of a Department of Innovation at Elf Aquitaine.

5. *Regulations.* Many leaders in industrial research management have written excellent articles on this most intractable of problems. We have not yet achieved a good balance between essential regulations, which protect the environment and our "quality of life," and those regulations which, in the name of protection, constrain our means of progress. The process of technology assessment must be designed to evaluate more clearly the positive, as well as the negative, effects of technology. Furthermore, any decision made by a regulatory body should embody a sound scientific basis and a clear assessment of economic impacts.

6. *Choice of technology for large investments in the public sector.* The renewed effort in basic technologies has at its other end the demonstration of new technologies. But such demonstrations are costly and risky. Few industrial companies are in a position to absorb such risks. However, the public sector continuously makes large investments in many diverse areas. For instance, the United States Government supports work on public buildings and space research; the French Government is involved in railways, energy, private housing, and other areas. It might be feasible to carry out demonstrations of new technologies related to the appropriate government-sponsored project, although the public sector cannot be expected to undertake all of the risk-taking investment either. In addition, we must have more reliable methods of evaluating risks (including sociological ones) and the importance of social experimentation in developing these methods must be fully emphasized.

7. *Representation of science and technology in all circles.* Scientific and technological information must be made available to all levels of society. Too frequently, however, scientists and technologists find themselves confined to separate professional "ghettos." The OECD Group of Experts recommends that this isolation be broken and that scientists and technologists be brought into collaboration with government. Their expertise is also sorely needed in industry, mass media, and legislative bodies. The results of the recent Heidrick and Struggles study are encouraging, since they indicate that chief R & D executives of large companies are becoming nearly as important as their financial counterparts.

8. *Consideration and acceptance of social problems.* This point cannot be stressed too much. The underestimation of public reaction to atomic energy was certainly costly in some of our countries; on the other hand, an overestimation of these reactions, leading to a delay of nuclear realization, could be equally costly. There are new situations basic to our

present society that deserve careful attention; new points of balance must be achieved between working hours and leisure time, and between quality of life and consumption of goods. Reflection on what might constitute a proper balance in each instance will lead to a new perspective on the kind of goods needed, and the processes to be used in their production.

The correct identification of specific problems for investigation in order to get a grasp on any of the aforementioned situations is enormously difficult. The development and correct use of social and economic sciences may offer the most reliable techniques for their identification. Technological assessment must not remain a theoretical exercise, but must become a dynamic, useful, detailed, and carefully investigated method of evaluation.

9. *Taking account of real needs: Is it possible? How to proceed?* This issue is clearly related to the preceding one. It has not yet been discussed by the OECD Group of Experts, but should be mentioned in this paper. The pertinent questions are: Is it possible to understand the real needs of people? If these needs are identified and fulfilled, will people feel grateful to the society that has provided for their needs? Or would new types of advertising be required to promote the desired feeling of gratitude?

These are certainly questions that call for further study, and universities should doubtless play a central part in their investigation.

10. *Education of our societies in technologies.* This last point reflects my own thoughts on education, which is now provided not only by schools and in families, but more and more by television and other media. Further great advances in telecommunications (involving electronics and space technologies) and methods of travel are imminent and indicate a central role for technology in shaping our world. Therefore, people from their earliest years should be taught to appreciate the realities and possibilities of new technological capabilities through educational programs in schools and universities and by means of information presented on television. Also, school curricula could be developed to teach responsible members of our society (other than scientists and technologists) something of specific technologies. Journalists, for instance, could be trained to understand technology more completely. Japan has been working in these directions and the success of their efforts should encourage us to try as well.

I would like to end this paper with one thought. The key to the solutions of our new and perplexing problems—solutions generally known but difficult to apply in our intricate societies—is surely the recognition that our countries and companies share similar problems, and that we would gain an infinite advantage by working together, rather than against each other, as we have in the past.

We must be confident about the future, and work constantly toward the fulfillment of its promise. We must also imbue the people of our countries with a vision of that promise.

A Perspective on Science and Technology Policy and Energy

JOHN C. SAWHILL

ONE CURRENT EXAMPLE of the impact of science and technology policy on all our lives was the call by Transportation Secretary Brock Adams to the nation's automobile industry to "reinvent" the automobile. This idea was originally laughed at by the manufacturers, but Congress recently staged a hearing on this theme, and now the automobile industry is already talking about joint government-industry efforts to redesign our most basic form of transportation.

The redesign of the automobile can be regarded as an inspired goal. However, confusion about the real problems concerning short- and long-term energy supplies causes most Americans to shut off their car radios whenever the energy problem is mentioned, but not voluntarily to use their cars less!

However, shutting the radio off and merely wishing for volunteerism—or Presidential calls for volunteerism—will not eliminate the problems we face this year and the years ahead. What we need immediately is a clear-cut national energy policy to deal effectively with the problems the United States faces for the next five to ten years, and for the next century.

Energy policy must be a continuum. The Israelis recognized this in their negotiations for a peace agreement when they asked for and received energy assurances for fifteen years. We, on the other hand, view the energy issue in terms of gasoline pump prices. Few Americans are prepared to discuss even next winter's problems, let alone the problems we will face in the near future.

In my view, a solution to the energy problem must be based on four cornerstones of national policy: (1) deregulating price; (2) resolving

John C. Sawhill, President of New York University and a former Administrator of the Federal Energy Administration, is on a leave of absence to serve as Deputy Secretary of Energy.

0077-8923/79/0334-0066$01.75/2 © 1979, NYAS

environmental-energy issues; (3) setting research and development priorities; and (4) developing an international perspective of energy supply and demand.

The Iranian situation clearly indicated that we must develop a comprehensive plan incorporation these four policies to accomplish two objectives: first, we need a plan which will enable us to minimize the risk of supply disruption and large-scale price increases during the next five to ten years while we will continue to be heavily dependent on Middle Eastern oil; and, second, we need a plan which will enable us to make a smooth transition over the longer term from liquid hydrocarbons to alternative, more available energy supplies.

Briefly, let us review these four points.

1. Decontrol of domestic oil prices:

The raising of domestic prices to the world level would encourage United States oil production, provide incentive for conservation, curtail oil imports and limit price increases on the part of the Organization of Petroleum Exporting Countries (OPEC), thereby helping to alleviate the pressures on our allies as well as those on our own economy. In fact, our current policy—a combination of price controls and the entitlements program—subsidizes imports and consequently provides economic incentives to expand, rather than contract them.

Moderate price rises now are the best protection against further sharp and destabilizing price increases in the future. Such price increases tend to expand both our supply and or demand options, and, to the extent that higher prices will be offset by efficiency improvements, they are not inflationary in the long run.

2. Environmental concerns:

Given that both environmental protection and energy development are important national goals, we must develop an efficient system for assessing the costs, benefits, risks, and trade-offs in the decision-making process. Without an efficient system for making decisions, time is lost, money is wasted, uncertainty is generated, and constructive action is thwarted.

Moreover, we must provide for an orderly process to eliminate duplication of effort among federal, state, and local government and to eliminate the conflicting requirements of various governmental agencies. In the absence of a resolution of these issues, industry cannot receive governmental assurance that the features it builds into plants today will not quickly become outmoded and have to be replaced tomorrow.

Further, the air pollution control program, as set forth in the Clean Air Act of 1970, and subsequently modified, defines national air quality objectives in a manner that is too narrow and absolute in light of the fundamental uncertainities and variabilities involved, and does not allow for desirable energy-clean-air trade-offs to be made.

Our current approach to air pollution is inappropriate in both its objectives and its methods of implementation. It imposes high energy costs with minimal environmental benefits, discourages innovation, and increases uncertainty and delay.

3. *Research and development priorities:*

First, the government should not become overly committed to building large demonstration projects for such energy projects as coal gasification and liquefaction. If these concepts are viable, the economies will be recognized by the private sector and it will make the necessary investments.

Second, priority should be given to unconventional and supplemental sources of fuel gas, such as "tight gas" formation, and other unconventional methane sources, all of which hold significant promise for meeting future national energy needs.

Additionally, the possibility of generating inexhaustible quantities of methane and/or hydrogen from solar and nuclear energy sources represents an important potential long-term United States option for maintaining and increasing the production and delivery of fuel gas. Therefore, research and development in these energy areas should be continued even though current price projections for their end use are high.

4. *International policy:*

The international implications of our domestic energy policy involve three major areas of concern.

First, under the International Energy Agreement, we are compelled to develop a domestic energy plan. As a signatory to this agreement, we are required to implement a demand reduction of seven percent if the International Energy Agency (IEA) of the OECD determines that a worldwide energy crisis has developed. And, "required" is the key word, for the IEA members can determine when an emergency situation exists, and this country must abide by that decision.

Second, we must develop a comprehensive hemispheric policy in full cooperation with Mexico and Canada. Mexico is the one known source of huge new oil and gas supplies, which are second only to Saudi Arabia's known reservoir of oil.

But Mexico's riches are a long way from being fully utilized. Surely, the time has now come for increased cooperation with and investment in Mexico. The problems of the past must be solved to the satisfacction of all parties. We cannot, and must not, attempt to dictate Mexican policy. So, too, should we work more closely with Canada, a country with large indigenous energy supplies, including a proven reserve of 60 billion barrels of oil. Working together with Mexico and Canada, we should actively encourage the development of a broad-based energy policy for the Hemisphere.

Finally, the role in the world energy situation of non-OPEC developing countries is frequently overlooked, although they represent over 40 percent of the world's population. However, although they currently require less than 10 percent of the world's commercial energy production, their commercial energy consumption is increasing rapidly. Certainly, as these countries develop their industrial base and strengthen their economies, they will present an opportunity to enhance considerably the future export markets for the United States. Even today, exports to developing countries have grown from $7 billion in 1960 to $40 billion in 1976.

The United Sates must encourage and support the efforts of these countries to develop indigenous energy resources, including oil, gas, coal, and hydroelectric power. These are countries that will expand United States export markets, possibly by a future energy source for us, and, minimally, with the development of their own energy supplies, may represent less of a drain on available oil supplies they become more industrialized.

Only by viewing energy usage within the context of these four policies can the United States hope effecitvely to manage its energy requirements for the years ahead. The time for merely talking about voluntary measures should come to an end, and a detailed plan of action must now be put into effect.

Food and Nutrition

SHERMAN K. REED

Strange as it may seem, no formal, broad, central statement of Federal policy exists with regard to food and agriculture. To discuss narrow "critical" issues, therefore, would be tantamount to putting the cart before the horse, or planning a political campaign without a party platform. Various national laws embody policy statements about narrow and specific agricultural goals, such as fair prices for farmers, or energy conservation on farms, but our implicit overall policy is simply to keep American farmers and consumers as contented, prosperous, and well-fed as possible. However, until an official, broad policy is forthcoming, any planning for the future, or "prioritizing" of "critical issues," is certain to be less than optimum.

For example, consider the following policy questions: What is the role of the United States in meeting pressures imposed by world population growth? How can agricultural production be used to offset the cost of imports and the trade challenge posed by oil-producing countries and foreign manufacturers? And, as regards recent developments in the cultivation of new plant species for food, pharmaceuticals and industrial purposes, what is their potential to relieve current and approaching shortages? These questions could be subsumed under a broad policy umbrella that could be called *the need to reduce the widening gap between world food supplies and requirements.* Similarly, a policy statement on the *need for cooperation and coordination between public and private sectors in agricultural research and development* would encompass such issues as assessing the benefits and risks of chemical additives in food and livestock feeds, or identifying new technologies and potential impacts in food production (for example symbiotic nitrogen fixation) and in marketing (for instance, the retortable pouch for packaging precooked foods).

In order that the many pressing issues in United States food and

---

Sherman K. Reed is Vice-President and Director for Chemical Technology of the FMC Corporation.

71

0077-8923/79/0334-0071$01.75/2 © 1979, NYAS

agriculture policy may be addressed systematically and effectively, a broader context for policy-making must be developed. The continuing absence of a policy-making framework based upon clearly articulated principles and objectives may well result in short-sighted, incremental policy decisions that will adversely affect the diverse utilization of our agricultural resources to their fullest potential. Therefore, let us consider the development of such a framework from the very fundamental premises upon which it would be based.

It is all too easy in America to forget the importance of food. We have such an abundance of it that we do not properly recognize what life is like without enough food. Without food, all other concerns vanish. National pride and ideology seem petty. Art, literature, and love become unreachable luxuries. Religion seems anachronistic. And even the sense of self dissolves. Everything must bow before the fact that the human body is a machine that needs at least 1,000 calories of energy each day merely to survive. There is a further fact: barely half the world's population of almost four and a half billion people get more than this minimum. Perhaps a third are at starvation level. We need to be reminded of this constantly. For we who live in countries with sufficient food are truly living in different worlds from those who don't.

As we talk policy, we might do well to think of the "haves" and the "have nots" as being on two different planets and think of the future in terms of "Star Wars" or some other popular movie in which the forces of Good fight the forces of Evil. And we must ask ourselves: Who are the forces of Evil? For each "have not," the answer clearly is: those of us who enjoy abundance, and do not share.

We are here to discuss and shape policy. It would seem well to agree on a definiton of policy. Policy, to me, is a process in which those who neither own nor directly manage a resource attempt to influence those who do.

If we look at our food-producing capacity as a total resource, it is obvious that no one manages it totally. But this does not mean it is not a *total* resource. This is not a particularly new thought, of course, but it is essential to the formation of policy. We have a finite potential; how well we manage the resource determines how close we come to the potential. And today, the policies, actions, and management decisions of those who neither own nor manage the physical resource increasingly determine the productivity of the total resource.

I shall return to this shortly. But let me say that the decision of a lawyer in the Environmental Protection Agency or a manager in the Food and Drug Administration or a chairman of a committee allocating funds for research are all as surely involved in the total management of the total resource as the chairman of the board or the stockholder or the researcher or the farmer or the implement manufacturer or the food processor, or all the other complex elements of our total food-producing capability. It is our collective policy decisions that determine our collective future and that of the people of the world.

One thing more before we plunge into specifics: policy, to me, means to reduce an intellectual position to its simplest, most direct form. Probably the greatest example of policy statements—a direct, simple, and powerful one—is the Ten Commandments, currently out of favor perhaps, but valid nonetheless. The commandment "Thou shalt not kill" makes no mention of forgiveness because one has grown up in a bad neighborhood, or is temporarily insane, or has acted in the defense of one's nation. "Thou shalt honor thy father and thy mother" says nothing about childhood trauma or about what to do with a mother who treats you like a child even though you are 59 years of age and a grandfather.

Can you imagine Moses staggering down from Mount Sinai with the Ten Commandments if they were written by a Congressional body, or a regulatory agency, or, alas, an international conference? He would have needed a 40-ton truck.

What I'm suggesting is that we stick with basic principles. The commandment to honor your father and mother does not attempt to tell you *how* to honor them; it tells you flatly to honor them. It assumes maturity. You have to work the details out for yourself. You know what's right and what's wrong. And if you don't, then you are the one to suffer.

Let me continue in the mode of Moses, for I want to present crisp statements of principle. I am made uncomfortable by the current effort to anticipate every conceivable act or action. This is what the existing plethora of law and regulation does, and it invites lawlessness by being unenforceable at best and capricious at worst. So let us stick with the most basic principles. Interpretations may vary. That's democratic, but the principle remains.

As a first policy statement, I suggest: Feed the hungry. If we cannot agree on this, then we will surely be a great distance apart on all other matters. None of us, walking out of this building today, would refuse food to a starving person outside the door. But that starving person—and hundreds of millions like him—exist right now as we discuss his and their fate.

If we agree to feed the hungry, then Policy II follows: make the land fruitful. There are those who think it obscene that by not optimally utilizing the land, we thus in effect reduce the total food resource while millions die of starvation each year. And I am one of those. Whatever your persuasion, if you agree to Policy I, then you will not find difficulty in agreeing to Policy II.

Of course, we live in a real world. There are economic matters and political concerns, the reality of the marketplace and of governments, to be accounted for. But I'm sure of one thing: The battle of people's minds always starts in their stomachs. Communism has always believed this. Lenin got the peasants behind him with cries of "Bread and land." Does someone faint with hunger want to hear about pollution?

This leads to Policy III: Thou shalt not dissipate thy nation's strength. If that sounds too Biblical, let me put it another way: When we reduce our nation's ability to be productive, we diminish ourselves. After World War II, we were so clearly dominant we could afford to be generous—with money, with technology, with expertise, with food.

Today, many nations have emerged with strengths that represent a considerable challenge: energy strengths, as with OPEC; people strengths, as with India and China and the overpopulated nations; and scientific and technological strengths, and manufacturing and marketing strengths.

In many areas, it is our food production strengths which keep us in the ball game. And don't underestimate them. Few of the oil-rich countries can grow enough food for their people. Is a barrel of oil worth a bushel of wheat?

Russia may glower and thump its chest alarmingly, but it needed the wheat it bought from us. Japan can't feed itself, and even less so, now that it can't sweep the seas clean of fish. Hourly, thousands of mouths are added to those already here. The scale of tomorrow's food problem will rival that of today's oil crisis. And I for one am perfectly willing to let food be a blue chip in the international game, especially when we are in a phase in which the dollar is battered and imports have decimated whole industries.

This brings me to Policy IV: Honor thy inventors and innovators. And might one not add, "that it go well with thee and thou livest long on the Earth"? What is this myopia that has led to a constant decline in invention and innovation for the past twenty years? There are those who wish to do away with the patent system, and thus drive out all incentive for inventive minds and venture funds. There are those who fantasize about some static, no-growth Utopia where everything is in balance. The yeastiness of new ideas, new products, and new ways disturbs that fantasy. To these people I can only say: There's no way back, particularly to a time that never was.

Science and technology have brought us this far. Science and technology must take us the rest of the way or we will all fall back into some nether place of lost dreams and unused opportunity, where the prime pastime will be deciding who was responsibile for the Fall.

Honor thy inventors and innovators—sweep their path free of even small obstructions instead of throwing up roadblocks that diminish their contributions. Without the creativity of our scientists and technologists we will grow weak in every way. Without it, for example, in the field of food and agriculture, our productivity will go down; our ability to feed the hungry will diminish; our land will not be fruitful; we will be a helpless giant. It has been my experience that the inventive individual has just about everything stacked against him or her. It took seventeen years to develop the jet engine; even the ballpoint pen took fifty years to be a market success from the time it was first conceived.

This brings me to Policy V: Let business be a "profit" in its own land. For many multinational companies, overseas business has long been subsidizing its domestic counterpart. There are many reasons for this, not the least of which has been the antagonism toward profit and toward business that seems to have emerged on the American scene. In part, it comes from those who do not understand, or more importantly, do not trust, a free market or free enterprise or private ownership. This is not the forum to make further inquiry as to why. But it is not the result of a plot, or of a foreign influence attempting a takeover. It is not even revolutionary. Rather, it stems, I believe, from the sincere desire of sincere people to solve problems of inequity and injustice and to right wrongs.

Incredibly, those who have done most to diminish freedom of action in the past twenty years are those who profess—sincerely, I believe—to wish to extend freedom. The rabbit-like increase of regulation has become so restrictive that America today is vastly different from that of even two decades ago. And it cannot continue.

Let me now state Policy VI: Approach improvement cautiously, for "better" is the enemy of "good." Almost all observers of institutions agree on one thing: If something works, leave it alone. More harm is done by tinkering with a functioning system than by an outright onslaught against it.

One of the great symbiotic relationships of the past century has been that of the government's relationship with private industry and with farmers in general in the United States. The national government, through land-grant colleges and the Department of Agriculture, has revolutionized the research and education process. No longer is knowledge passed along only from father to son. Advances in the laboratory reach the fields in an incredibly short time. The government has supported great advances in hybridization in plants and animals. It has helped the farmer understand his land and how to use it, and how to keep it from eroding or blowing away. The government has educated consumers and food processors and has created a climate in which individual and corporate entrepreneurs could confidently invest in research and development and innovation in every area connected with farm implements and processing skills, packaging and distribution, chemicals for food nutrients and insecticide, rodent and weed control, and for soil conditioning and growth—and it could be confident that resulting advances would be greeted intelligently and enthusiastically.

The relationship has not been perfect, *but it has been good.* This is one of the reasons we have a food production capacity such as the world has never seen. And I don't know anyone who can tell you exactly and specifically why it has worked. But it is obvious that it has succeeded according the principle that some things naturally should be done by government, and the rest should be the responsibility of the dynamics of a free market and of private enterprise and ownership. I am fearful that many well-intentioned people will attempt to make better what is already good, and so achieve neither their objective nor mine.

This brings me to Policy VII: Anything that can be accomplished by private industry and private enterprise, should be. And I mean *anything.* Does this sound radical? I believe firmly that there is no better regulator than a free market. If it does nothing else, it allows the consumer to vote daily by means of his purchase. The free market is democracy at its best. And it is unsentimental. If you do not understand the public, you do not survive. This is all right. The entrepreneur has yet to become an endangered species.

"Anything that can be accomplished by private industry and private enterprise should be:" does that sound monopolistic? Not at all. For there are great segments of need which only the public sector can fulfill. Except in rare situations, business has never been able consistently to pursue basic research, and much has yet to be learned in the areas of biological processes and plant sciences, in environment protection and food safety, in world food requirements, soil and water conservation, pollution, weather, climate, aquaculture, silviculture, post-harvest technology— the list is long, and human nutrition should certainly be at the top of it. We know more about the nutrition of chickens and hogs than we do about that of humans. I can think of no better area for investment of public funds than in those areas where private research and development cannot go.

But there is a danger. The government, in an effort to be responsive to all the needs of all the people, all too often spreads its resources so thinly that nothing important gets the attention it deserves.

And so, let me state Policy VIII: Waste not thy substance on myriad concerns. The attempt to do something for everybody, to attack all problems, dissipates our strengths and produces trivial results. We have long entertained a rather naive belief that if we throw enough money at a problem we will find a solution. Well, certainly, the money is important, and the people too. But there isn't enough money to go around to solve all problems simultaneously. And there are certainly not enough good people. Funds and effort are dissipated on trivial concerns, or worse, an important area of research is understaffed and underfunded. Let us pick four or five or six *major* areas of investigation and conduct research that will benefit a broad spectrum of people.

I have already mentioned nutrition. Other issues might range from new concepts of land use to new sources of nutrition and how to get new foods accepted. For it is obvious by now that if we are to feed the billions, we are going to need to utilize every food-producing method conceivable. I do know we will never arrive at major solutions if we continue do dissipate our energies and skills and resources in too many directions. And we can accomplish our goal only if we recognize that we are all involved in managing the same total resource.

This brings me to Policy IX: Be not arrogant in the uses of power. Probably the most difficult job in the world is to use power wisely. And one might regard business and government today as two opposing power sources.

Some of the sources of friction are attributable to business itself, because business has not always been ready to recognize that ownership of a national resource is both a privilege and a responsibility. To mine, to refine, to grow, to process that which comes from our lands, puts business in the position of a kind of public surrogate. For no one can readily *own* the land or the air or the water or the other natural resources. The public has rights which transcend the title of ownership. And because the wealth of the nation depends on its resources, it follows that no one—individual, corporation or government agency—has the right to squander those resources. In the pell-mell pursuit of profit, one cannot casually befoul the air and waters or lay waste the land. In the sometimes desperate day-to-day battle for survival as a business entity, a private enterprise cannot callously sacrifice the minds and bodies and health of both workers and the general public. The public will not be damned. The public will be heard. The majority does rule, even if it takes time for it to be heard. So the private sector cannot claim that ownership is all and public rights should be subordinated. That is an arrogant use of power.

On the other hand, the emerging power centers in the public sector—and I refer primarily to regulatory agencies—must not take a narrow, arrogant view, or regulate capriciously. One reason for this pitfall is that regulation is primarily negative in nature. One inevitably thinks of the taboo cultures of more primitive societies. The taboos become so all-pervasive that little free choice is left. A society grown static and uninventive dissipates its energy in avoidance. Creativity dies. Be not arrogant in the uses of power either as owner or regulator, for as noted before, we are all involved in managing the same total resource.

This brings me to Policy X: Be mindful of the fragility of human institutions. Democracy is fragile. Freedom is fragile. A free market is fragile. Business, industry, and the entire agricultural establishment are fragile.

Thousands of farmers get forced out of business each year. I don't know a business in this country which could not be in trouble in five years and gone in ten if the wrong conditions came about. The death of businesses, or of farms, or of processors and distributors, is frequently hidden from public view by merger.

Occasionally, as in my home town of Philadelphia, we see a huge supermarket chain go under. A bank here and there succumbs, or a chain of department stores like Grant's. But 50 percent of small businesses fail. And I am told by experts that increased regulatory restrictions will almost certainly doom smaller concerns—and I include those even in the 2- to 300-million dollar range. Too much time, effort, and money are being spent in nonproductive ways.

When I suggest that we be mindful of the fragility of our institutions, I am not calling for handouts. Truly, I am asking for the free play of economic and market forces wherever possible. I deplore the fact that the business and agriculture communities want support rather than the operation of a competitive market. I think that such support is bad for everyone. It has been called socialism for business, and it is debilitating.

On the other hand, I think that knowledge gained through the public expenditure of research funds should be turned over to the private sector without a moment's hesitation. Government should not get involved in areas where the business sector can function better. But providing knowledge in this way would not be socialism; it would be an investment, and a wise one at that. Because, for every dollar of profit earned in the business sector, at least 50 percent will come back in the form of taxes in one way or another. It is an investment which has to pay off, whether in increased jobs or in increased production. It is a specific example of my belief that we must manage our total resources as a whole.

So let me end with a seeming redundancy by way of emphasis.

1. American food and agriculture resources must be viewed totally, as a total resource.

2. Each of us must function in terms of that total resource.

3. If we do, we can alleviate the friction and diminish the myopia that hamper our efforts.

4. As we accomplish this aim, we come closer to making our total resource totally productive.

If we do not, History will deal harshly with us. For it is indeed obscene to overeat while others slowly starve. And we must assume that we are the enemy if tomorrow will be no more than a copy of today.

Overview of Policy Issues: Panel Report

Panel Chair: GRACE L. OSTENSO
Panel Members: Reynald Bonmati, Richard H. Braunlich,
R. W. Diehl, Felix J. Germino, Steven Marcus,
Andrew H. Pettifor, Geoffrey Place, and
Monte C. Throdahl
Rapporteur: Susan M. Leech

THE FOOD AND HEALTH PANEL* was concerned with broad areas of relationship among agriculture, food, nutrition, and health as a continuous process. It was agreed that the problem of feeding hungry people is of prime concern and that the demands and constraints of this task call for careful and comprehensive consideration.

Actions, regulations, and management decisions are being made that increasingly determine production, distribution, and use of food and agricultural products. However, broad policy guidelines, as such, have not been established. At the same time, government regulation has become pervasive, inhibiting, and capricious. The purpose, process, and benefits need therefore to be carefully re-examined.

The agricultural sector is presently heavily regulated, from the research stage to the final production. Government funding has produced an impressive and comprehensive system of land-grant colleges and agricultural experiment stations that perform valuable research and testing. But, because of the lack of overall policy guidelines, the majority of the research projects have not proceeded beyond testing. Results gained through federally funded research are disseminated through a variety of channels, but there is a need for a comprehensive research repository and retrieval system readily available to the private sector. Additional research and increased cooperation and exchange of viewpoints among

---

* The majority of the panel members able to attend were industry representatives and thus the discussion predominantly reflected the viewpoint of industry rather than the perspective of the regulator and consumer.

Grace L. Ostenso is a Science Consultant with the Subcommittee on Science, Research and Technology of the Committee on Science and Technology, United States House of Representatives.

0077-8923/79/0334-0080$01.75/2 © 1979, NYAS

government, industry, and academia are necessary to develop a sound basis for regulatory decisions.

Agricultural production is heavily regulated by various government agencies, each acting within their specific area of concern. Production is becoming more and more constrained as producers comply with a myriad of regulations which appear to lack the guidance of a coordinated policy. Often not enough knowledge exists and not enough basic research is performed either to warrant or to disprove the need for regulation. Moreover, the cost of compliance with regulations is becoming increasingly prohibitive; time lapses caused by paper work and periods of waiting for response add to the costs of compliance. Policy research is needed to determine techniques for increasing the efficiency of the regulatory process and feasible alternatives to regulation.

Alternatively, food products and the food industry presently are subject to minimal regulation. There is a fear, however, that regulation will increase to the point that product formulation may be constrained by government requirements to provide more specific information on nutrient and ingredient content. Steps in this direction have been taken by means of government attempts to develop nutrient profiles for processed foods to achieve comparable nutrient content. However, there is a paucity of basic research, a small science base, and inadequate knowledge of nutrition upon which to base requirements for most food products. Industry and universities may be in a position to undertake the necessary research to develop the data base for effective regulatory decisions and the techniques necessary to reduce compliance costs.

Basic research requires a massive effort and substantial start-up costs. Policy to direct basic research objectives and related investment is desperately needed. For example, patent law provides an obstacle for industry to conduct basic research since the protection afforded by patents does not allow adequate time to do basic research and conduct necessary testing and marketing experiments. Industry is also concerned with the problem of expending large amounts of resources on food processing and packaging without reasonable assurance that the resulting product will not be prohibited or that regulatory guidelines will not be changed prior to completion of the product development process. Government, then, has set up built-in disincentives for industry to take risks. Moreover, industry is faced with a lack of key trained personnel for basic research. Incentives could be provided to universities to train personnel with the necessary expertise to meet industry and government needs in developing food policy, conducting research and assessing regulation. Efforts should be made to involve industry, government, and university in seeking solutions to the issues confronting effective and efficient food regulations. To avoid serious problems concerning food supply, regulation, and costs the most pressing needs are for policy guidelines, basic research, and trained personnel.

In developing food and health policies, we must consider how food is used as a resource, the multiple impacts of any policies formulated, and which policies would best encourage development of our potential food resources to meet nutrition and health goals. To address this effort, various components of the food system were reviewed and conclusions were drawn highlighting the critical areas of need.

The various components of the agriculture, food, and nutrition system were determined to be: seed and fertilizer, land, harvest, transport, processing and preserving, packaging, consumption, and general welfare (health). Labor and machines function as primary inputs and the government's impact on the entire system is in the form of regulation. At present, government regulation affects most directly the first part of the system through rules laid down by the Environmental Protection Agency, Department of Agriculture, and Food and Drug Administration. Increased regulatory involvement in the consumption area of the system is anticipated. High technology, a major resource of American and modern industry, clearly pays off through abundant harvests. From the perspective of financial returns on such harvests, however, the food system becomes so disaggregated that high technology usage and innovations are no longer cost-efficient.

This then identifies a major opportunity for research and development and a challenge to initiate and develop more effective science and technology policy. The total cost of food is composed of incremental costs associated with every phase of production—from cultivation through final purchase at retail. The efficiency of technical applications and the contribution by government have improved the cultivation stage of food production. Improvements in the total system, which will then benefit the consumer, require creative policies to provide incentives to address directly problems related to the disaggregated nature of the system.

With the challenge to encourage development of our food resource potential, the major areas of concern are: distribution and bulk preservation of food; processing and preserving to meet general nutrition and food safety requirements; cost-efficient operations in the food system; determination of food effects on health status; and the availability of trained personnel to address these challenges. Several approaches may be needed. The primary requirement is the establishment of a scientific and technological base that can identify specific measures of health status that have a relationship to nutrition, and measures of nutritional status that can be correlated with food consumption patterns. A secondary requirement is to educate the public about the relationship of nutrition and health and to provide the consumer with an effective nutrition information system to facilitate informed food choice.

Institutions involved in food and health may be ill-prepared to undertake this work. It seems likely that industry is better able to respond to the need for food composition data than either academia or government. However, research in related disciplines is not adequate and universities are underinvested in research and training areas. University and industry cooperation could help to ease the burdens of these challenges.

The greatest opportunity for development of policy rests with the government. If government places a higher priority on food and health, it can lead the way in enabling scientific and technological development of our food resource potential. Government's primary role is to provide incentives through funding and new patent and licensing arrangements for private industry to innovate, research, and develop. Food and health policies must be developed with the cooperation of the public, industry, academia, and government and greater emphasis must be placed on establishing goals rather than stipulating means and on allowing the marketplace to operate with greater flexibility.

The first and immediate challenge before us, then, calls for a meeting of industry, academia, and government to look at problems and needs and to establish coordinated nutrition, food and agriculture policies. Industry should seek every opportunity to present testimony and should strive to involve the public in this effort to solve our food and health problems. Congress should devote greater attention to these issues through agriculture, science and technology, health, and budget committees. Universities, too, can play a role in the provision of research and data, personnel, and expertise in bringing these problems to the attention of the public and the government. Industry, university, and government must make every effort to expand the ongoing dialogue. Developing our food resource potential is a complex task, but it is certain that the technological and scientific requirements can be developed if government, industry, and university cooperate in forming a more coordinated food and health policy.

Materials

WALTER R. HIBBARD, JR.

A REVIEW OF RECENT policy studies relating to materials and minerals suggests that there are two predominant issues: (1) The need to create a cooperative environment between government and industry in order to assure a strong future in the field of materials. (2) The deterioration of long-range research and the emphasis on short-term development that have been brought on by profit squeeze, energy costs, and environmental and regulatory restrictions. The symptoms of these maladies have been studied and reported in detail, but the disease, namely, a national policy of protecting society from industrial excesses, thoughtlessness, or environmental and social neglect by means of regulations, must be directly addressed.

Government policy consists of actions such as legislation and regulation which are implemented and enforced. Therefore, statements of policy, without action, are not policy. For example, despite declarations of government officials to the contrary, the effect of United States energy policy has been to increase the use of imported oil and gas by regulation, to curtail the use of coal and nuclear energy sources by regulation, and to blame any problems that arise on industry. The enforcing persons play key roles in policy implementation.

Congress, in response to single-issue lobbying dramatized by the media, has legislated controls and regulations over every aspect of the materials cycle as follows:

Mining:	Mine Health and Safety Act of 1977
	Surface Mining Control and Reclamation Act of 1977
	Land Policy and Management Act of 1976
Refining:	Clean Air Act of 1970 (amended 1977)
Processing:	Water Pollution Control Act of 1972 (amended 1977)
	National Environmental Policy Act of 1975
Design:	Design Liability actions
Manufacture:	Occupational Health and Safety Act of 1970
	National Energy Act of 1978
Assembly:	Power Plant and Industrial Fuel Use Act of 1978

Walter R. Hibbard, Jr., is University Distinguished Professor of Engineering at Virginia Polytechnic Institute and State University.

0077-8923/79/0334-0084$01.75/2 © 1979, NYAS

Use:	Consumer Product Safety Act of 1973
Recycling:	Resource Recovery Act of 1970
	Resource Conservation and Recovery Act of 1976
	Energy Tax Act of 1978
Disposal:	Solid Waste Disposal Act of 1965
	Toxic Substances Control Act of 1976

Serious concern for the depressed condition of the materials industry in the United States has been expressed in numerous reports; several major reports are listed in the Appendix at the end of this article.

The regulatory laws have led to the proliferation of regulations by avid bureaucrats. But concerns for the conserving of materials have led merely to studies and reports with no policy implementation. As a result, regulations are shaping the destiny of the materials industry in the United States economy.

Regulations and their enforcement have substantially affected energy aspects of the materials cycle from production, to fabrication, to use, and finally to disposal. For example, regulations have caused severely inflated costs. They have added nearly $5 billion to the Federal budget of 55 agencies whose 126,000 regulators force the expenditure of $102 billion per year by industry for compliance and information. Compliance with pollution regulation requires expenditures of $47.6 billion per year, of which air pollution accounts for $13.1 billion per year. More positively, however, benefits arising from air-pollution control are estimated to have saved $22 billion in costs of health care, property damage and harm to fish and wildlife.

Regulation also directly affects exploration and supply; it has halted the expansion of United States basic-materials capability and led to increasing imports; it has had an impact on the recycling process; it has generated substitution incentives by limiting access to sources of scarce materials; and it has imposed conservation measures and dictated energy use.

In addition, there is an enforcer-violator type of situation where the enforcer changes the rules periodically and serves as both judge and jury in cases of alleged infraction. This situation subsequently leads to an adversary relationship that generates an environment of suspicion and distrust, and deeply erodes the possibilities for cooperation between industry and government.

Regulation now anticipates and legislates new technology before it exists. Scrubbers and catalytic afterburners are examples. Nearly one-third of United States industrial research and development (R&D) is estimated to be involved in some way in reducing the costs imposed by regulation.

As a result, we are now entering an age of little new-materials R&D of a long-term innovative sort, particularly in the area of new processing discoveries.

As an example, the passenger automobile is regulated to be safer, less polluting, more energy-efficient and more durable within a certain time limit As a result, R&D in the auto industry is concentrating on these factors (which are not mutually self-consistent) and this has caused United States companies to manufacture overseas with foreign technology, and to import into the United States.

The basic materials industries are also trying to modify their existing facilities to satisfy regulatory guidelines while expanding their operations overseas to produce basic materials and, sometimes, manufacture products and components for importation into the United States.

In general, then, R&D has been forced to focus on short-term development of solutions to the problems created by profit squeeze, regulation, and energy use. Even university research has responded to funding directed at environmental protection and energy conservation. And as a result, long-range research has deteriorated.

Science and technology policy probably cannot rectify this situation, since existing antimaterials policy stems primarily from grass-roots, single-issue laws. Congress is likely to change these laws only under pressure, and voters will press for change only when their quality of life seems threatened.

Materials Issues

Materials issues relate to a variety of needs in our society. We must provide an adequate supply of basic materials by reducing imports or assuring their availability. We must increase domestic productivity. We must minimize negative environmental impacts and energy costs, and we must conserve and recycle materials. At the same time that we seek to fulfill these needs, we must respond to government regulations that often seem at cross purposes with economic goals.

United States basic-materials industries are expanding overseas where there are more incentives for productivity and there is greater government cooperation. Aluminum imports have increased by more than 100 percent since 1975, and these are largely used by United States industry. Concurrently, imports of refined copper have increased by 242 percent, and iron and steel by 44 percent. Imports of raw and processed materials have increased to $20 billion per year. The net value of imports relative to exports has grown from approximately zero to $6 billion per year.

Although there is a world surplus of the aforementioned materials, user industries seek lowest prices with little concern for an ailing domestic minerals industry. They argue that domestic resources are conserved and pollution is avoided by not mining and refining those materials in the United States. However, this policy has seriously affected employment in the industry. Between 1975 and 1978, unemployment increased by 1,000 workers in the aluminum industry, by 4,500 in the copper industry, and by 55,000 in the iron and steel industry.

In the basic materials, certain overseas locations offer comparative advantages relative to those of the United States. There are richer ores, cheaper energy, lower labor costs and, in some instances, tax advantages and ready capital access. In general, pollution-control regulations are less stringent and safety regulations are more compatible with standards of industrial procedures. In fact, except for copper, many of the overseas resources are owned or operated by United States corporations that are importing into the United States market as well as into Europe and Japan. Overseas processing is expanding into the primary processing of shapes at the source. The prospect of a major imbalance in the domestic capacity of metals-producing industries (relative to demand) over the next decade appears almost certain, because existing United States facilities may become economically unprofitable.

In addition, American companies are gradually stepping up overseas manufacture of consumer goods and components for machinery and electronics using foreign materials and then importing them into domestic markets. This suggests that American goods are no longer cost-competitive.

An even higher percentage of basic materials being imported both in their primary state and in consumer products implies both positive and negative effects. The resulting manufacturing costs in the United States are lower because of richer ores, lower labor costs, and regulations that are less restrictive and less costly to comply with. Less pollution is created at home and energy resources are conserved. On the negative side, however, domestic employment in these industries and their supporting services will decrease. Adverse pressure on the balance of payments will continue and uncertainties about supply reliability will increase.

By contrast, Germany and Japan thrive on imported materials. But in those countries industry has a more congenial relationship with government, the legislature, the media, and thus the people. Moreover, these governments assume a more direct role in assuring supplies and maintaining balances.

ENERGY AND ENVIRONMENT

Materials production is energy-intensive, and necessitates major programs of energy conservation in its industries. Because of the cost and quantities of energy used, materials fabricators and product manufacturers are designing products with materials and processes that reduce energy consumption while increasing productivity. For example, an energy-conserving set of processing, forming, and fabrication operations would involve lighter weight and more unitized construction (to reduce energy in joining and assembly).

Moreover, the development of suitable materials for energy-generating and energy-conversion systems is critical in several cases, such as coal conversion and solar-electric power, and may be the rate-limiting factor in their commercial feasibility. Such energy-conserving measures, however, often confront environmental restrictions that frequently call for an increase rather than a reduction in energy consumption. Thus, seeking to achieve both energy conservation and compliance with environmental regulations poses a considerable challenge for materials R&D.

RECYCLING

Materials recycling is encouraged by several Federal acts but is limited in practice to those situations where a cost advantage exists. Manufacturers such as Western Electric, General Motors and General Electric are recycing significant amounts of materials internally. More designs simplifying recycling are needed and may emerge where profitable.

Recycling to minimize energy requirements and to minimize imports can be cost-effective, and regulations may require it. The Resource Conservation and Recovery Act of 1976 requires that all waste materials be stored in a sanitary land fill and that open dumping cease and its effects be repaired. This law provides incentives for the recycling of materials, particularly municipal waste. The technology is available to accomplish this objective, but institutional, capital-formation, and waste-collection problems exist. Again, technology leading to improved design of recycling may stimulate this effort. Ease of identification of recyclable materials is essential, as exemplified by the all-aluminum beverage can and the lead storage battery. However, changing the trends of our throwaway packaging is unlikely. Nonreturnable containers with convenience features will probably persist as we continue to generate solid waste. Technology is needed to sort the waste so that it can be recycled in its most useful form. The concept of a waste dump as a man-made resource is intriguing. Most municipal waste may be a useful source of copper, gold, silver and other materials, once the burnables are consumed and the iron, aluminum and glass is removed. Such an approach would be worthy of policy consideration. In the case of hazardous materials which cannot be recycled, intensive study is required to evaluate risks and to find how to neutralize or contain them in a rational, acceptable system.

Conservation

Although at present there may not be a world-wide shortage of materials, it is anticipated that such shortages may develop with respect to selected, critical materials since new supply is not keeping pace with demand. Imports of materials by American firms indicate that price is an important factor, for overseas production costs are generally lower than domestic costs. In addition to cost as a primary motivating factor is conservation. Weight reduction and fabricability are important where energy is a factor. Experience with attempts at fuel conservation has shown that conservation will not be effective unless there are shortages and associated crises.

Conservation of manufacturing materials can result from the use of more durable materials. If, for example, American-made automobiles lasted as long as European-made ones, fewer materials would be required to provide the same functional service. Corrosion has been estimated to cost the American consumer $70 billion annually. In general, technology is available to minimize corrosion, but at a cost that the consumer is apparently not willing to pay. Design for maintenance, servicing and repair has given way to design for replaceable components. There is an urgent need for greater awareness of the cost-effectiveness of maintenance and repair, particularly in view of the impending shortages of some materials and their associated cost increases. New technology is needed to stress designs that incorporate functional materials that do not corrode or wear out and that permit protection and necessary servicing by the owner or maintenance man.

Substitution and Supply

A national policy on materials substitution is lacking. It is clear, and it has been amply demonstrated in our economy, that materials substitution can lead to advantages such as improved fabricability, lower weight, ease of recycling, improved joining and assembly, improved cor-rosion resistance, improved wear resistance, improved responsiveness of availability to abundance or scarcity, and lower cost. The supply of materials is subject to political factors both here and abroad, and to costs, scarcity, regulation and public opinion. Thus, there is an urgent need for a system that can give early warning of impending shortages in the supply of materials, and that can provide for an orderly transition to substitutes without economic dislocation.

Research and Development

Recent studies indicate that less than one-third of total R&D expenditures is devoted to materials. Current materials R&D is mostly concerned with regulation, antipollution measures, health, safety, energy conservation, and so forth. Little effort is dedicated to the development of new materials, or to durability, availability and the other concerns expressed in this article. As a result, productivity is not receiving its historic R&D impetus.

The most exciting R&D opportunities lie in treating the entire materials cycle as a system. For each stage, a materials objective might be devised as follows: Production should minimize pollution and energy use. Fabrication should be designed for unitized construction. Product assembly should be designed for disassembly and sorting to facilitate recycling. Maintenance may be reduced by designing for durability. (Note that the question of new materials is not the issue. Considerations relating to materials processing and assembly are predominant.)

To achieve these goals, materials R&D must re-emerge under the stimulus of technical opportunity as it did twenty years ago. Science and technology policy can affect this regeneration.

Summary

In summary, key general issues related to materials are:
- Lagging innovation and productivity.
- R&D emphasis on short-term incremental improvements.
- Need for improved cooperation between government and industry.

More specifically, materials issues relate to these needs: (1) alleviation of increasing energy and environmental costs; (2) development of new materials to commercialize new energy processes; (3) more thorough assessment of the risks of hazardous materials and how to contain or neutralize them; (4) conservation of materials by design, substitution and recycling, and provision of alternatives to imports; (5) early identification of supply and availability problems; and (6) recognition of political factors in supply, overseas shifts, regulations, and public opinion.

In light of this overview of issues in the materials area, let me pose several questions on the relationship of science and technology policy and the materials industry:

1. Do we need a national materials policy?

2. How can long-range research and innovation be restimulated?

3. Would the materials industry thrive under a Department of Natural Resources?

4. Should regulations be reviewed in advance to evaluate the science and technology upon which they are based?

5. Is a materials impact study needed with each new regulation or ruling?

6. Is science and technology policy an effective antidote for other national policies that are draining the materials capability of the United States?

7. How can materials issues best be integrated into the current cabinet-level studies on nonfuel minerals policy and on innovation?

8. Is there a recognizable science and technology policy relating to materials?

These questions have no simple answers. They do, however, indicate the range of priorities for policy development and evaluation over the next several years. Because of the complexity of the issues, the magnitude of such an effort is awesome. But is is essential to the overall health of our economy and quality of life.

APPENDIX: LIST OF RECENT ARTICLES FOCUSING ON DEPRESSED
CONDITION OF MATERIALS IN THE UNITED STATES

1. Materials Policy Commission Report of 1952.
2. National Materials Policy Act of 1970.
3. Mining and Minerals Policy Act of 1970.
4. NAS/NRC/NAE Report of 1972: Elements of National Materials Policy.
5. NAS/NRC/NAE Report of 1973: Man, Materials and the Environment.
6. NCMP Report of 1973: Materials Needs and the Environment, Today and Tomorrow.
7. NAS Report of 1973: Materials & Man's Needs. Materials Science and Engineering (COSMAT).
8. Report of National Commission on Supplies and Shortages of 1977.
9. Five Henniker Conferences on National Materials Policy.
10. OTA, GAO, CRS studies of materials.
11. Several hearings before Congressional Committees on National Materials Policy Bills.

Overview of Policy Issues: Panel Report

Panel Chair: N. BRUCE HANNAY
Panel Members: J. C. Agarwal, Alfred E. Brown,
Kenneth Gordon, John Holmfeld, Lionel Johns,
Ursula Kruse-Vaucienne, Charles F. Larson,
Mary Mogee, A. E. Pannenborg, S. Victor Radcliffe,
John M. Rozett, Allen S. Russell, Bruce Smith,
Lowell W. Steele, and Max L. Williams
Rapporteur: Arthur Norberg

THERE ARE TWO FRAMEWORKS that might be useful in considering materials policy issues. One would be a uniquely materials-oriented framework that arranged issues according to their relation to availability, processing, use, and disposal. The other would be an issues-oriented framework that highlighted issues of international, national, and regional character; these might include short- and long-term national security and international economics issues; issues requiring multilateral, bilateral, and unilateral agreements; and United States national and regional issues. Whereas the former framework focuses on the materials aspect of the issues, the latter is possibly more useful in eliciting the attention of the United States Congress and other policymakers.

There is currently intense interest in the availability of materials, especially those mainly imported by the United States. The consensus of expert opinion is that, despite forecasts of doom, worldwide shortages are not likely in the next several decades. Most, and perhaps almost all, critical materials supplies worldwide are adequate to meet anticipated demand. Worries about natural limits might exist 50 to 100 years from now. For the United States, there are three concerns about availability. First, there is the question of political access. Dislocations could occur due to political problems or the formation of cartels. Past experience indicates that both of these events would probably be temporary and reversible.

Second, there is the question of price. Up to 1973, the demand for ores

---

N. Bruce Hannay is Vice-President, Research and Patents, of Bell Telephone Laboratories and a former President of the Industrial Research Institute.

0077-8923/79/0334-0092$01.75/2 © 1979, NYAS

and minerals by United States industry kept at least ahead of the demand for the resulting finished products. The increasing demand for products in the 1970s by expanding economies around the world pressured prices of materials upward, and, simultaneously, the materials-producing industry found itself lacking capacity to handle the demand. When the economies slowed down after 1973, the prices of materials decreased. In any predictions of future supply and demand the question of availability must involve a judgment whether the initial price rise or the later price decline was the more significant indicator of the future. There are some similarities to the recent issue of the availability of energy raw materials, and it remains an open question as to whether policies will be developed to manage nonfuel resources better than we have managed fuel resources. The opportunity exists, but whether any action will be taken in the absence of a perceived crisis is unpredictable.

Third, and needing more study, will be the ability of American industry continually to alter processing capacity to meet demand. At present, excess capacity exists, but various factors including capital costs and regulation have lengthened the response time to add needed capacity. A lack of capacity, of course, can intensify the problem of tightening supplies on a short-term basis.

It is worthwhile to ask whether anything can be learned from a consideration of recent discussions having to do with energy policy. What has emerged from these discussions is a reasonable consensus that there are short- and long-term issues. The long-term issue, which started out being the major issue, was: Given the decline of world oil resources, how should the United States undertake the transition to a new energy-resource mix? But the short-term issues dominate at the moment, and suggest the following questions: (1) What steps should the United States take to solve national security problems associated with a high reliance on a major imported fuel (namely, oil)? (2) What steps should be taken to deal with the imbalance of payments with respect to that dependence? Similar issues could arise if the availability of nonfuel materials was suddenly impaired. It seems apparent that resource questions must be described in a compelling manner to capture national attention.

This consideration can be expanded through an examination of issues faced by the copper industry. There is a need for continued availability of copper at reasonable prices in the American economy. One problem is that existing indigenous supplies no longer are economic to develop—such as most of the copper in the State of Arizona—because of other priorities that public policy has put in place, largely through actions of the kind brought about by the National Environmental Policy

Act. This issue, then, resolves into two policy issues, namely, foreign and domestic. One of the foreign policy questions raised is: Should the United States Government deliberately encourage the active participation of American copper companies in the exploitation of overseas resources in order to develop a more reliable continuous supply of copper to the United States market in the future? Similar questions are possible for other industries as well. A resolution of this issue might arise from a larger policy context and follow from any national policy on balance of trade. This suggests the need for a national policy on whether or not we want to be internally self-sufficient in the production of materials, or whether we want to support under- and undeveloped nations through imports.

Issues on the domestic side, and more specifically, science and technology policy issues, are: Should the government be looking into the future, when richer ore supplies in other countries are going to be exhausted, and seek to assure the development of new technologies that will allow economical exploitation of indigenous supplies of ore? A somewhat larger domestic issue relates to the provision of economic opportunities for people displaced from domestic materials-producing industries if national policies result in decreased activity in these industries.

Before addressing such issues, we should examine present policy to discover its weaknesses and inadequacies, and could, with regard to availability, do so by raising the following considerations:

(1) A national and international inventory of materials is needed. Something is known about such an inventory in terms of certain reserves, but about total resources there is a high degree of uncertainty. From a knowledge of the amount of total reserves and the rate of their depletion, an appreciation of trends will emerge that might indicate actions needed concerning availability, both nationally and internationally. Particularly important would be indications in the trends that declining availability and the lack of substitutability will create economic, political, or social dislocations. The implications here for import policy are obvious.

(2) Emergency conditions created by such dislocations might be mitigated by measures such as stockpiling, or by economic measures.

(3) At the same time, a better appreciation of the rate at which minerals are discovered and processed will reveal any impediments to discovery created by present policy.

(4) Another area related to inventory and to discovery rate is the question of access to public lands.

(5) Once an understanding of resources is achieved, the question of

materials substitution arises. What substitution research and development is needed, and should we have a policy to guide it?

(6) On a broader plane, there is the matter of American policy approaches to international development of the resource potential that lies in both the developed and lesser-developed countries. Should the United States be involved in promoting this development? Are unilateral, bilateral, or multilateral agreements desirable to promote a policy relating to the discovery of resources, particularly short resources, in areas of the world where exploitation has not occurred?

In sum, most policy concerns about materials availability are contained in larger policy contexts. For example, a United States policy to lessen dependence on imports is part of the balance of trade issue. And what are the roles of the public and private sectors with respect to materials availability? Materials issues are embedded in policy issues involving the relations of the United States with other governments—economic, military, and political—and are involved as well in a vast array of domestic policy issues of broad scope.

If supplies of certain lower-cost materials become unavailable, either through discontinuities in events abroad or through a domestic policy of promotion or restriction, then there will be increased need for attention to problems of the use, recycling, and disposal of materials, as well as to all phases of minerals extraction and processing. Questions relating to materials substitution will arise, including use of renewable resources as substitutes for nonrenewable resources to provide materials for manufacturing. The gap between technologies for processing sophisticated materials (for example, semiconductors) and less exotic materials (steel, for instance) will focus attention on deficiencies of knowledge for the latter classes.

In these areas science and technology, through research and development, can have a strong impact, and new policies for their prosecution may be required. Basic knowledge in the mineral sciences, for example, in the area of the thermal properties of ores and the processing of ores, is far short of what would be needed; greater knowledge at a basic level would lead industries to more efficient use and processing of minerals. Research on ways to employ renewable resources as generators of materials appears to be off to a promising start, and the large-scale harvesting of renewable resources (such as forests) has received policy attention. However, controversies still rage about basic scientific information on the best harvesting techniques. As another example in the resource area, proper use of satellite data of the earth's surface to inventory natural resources requires much refinement of correlation techniques to match the data with surface phenomena. Equally promising opportunities for science and technology exist in materials processing and use.

Present concern for the use of natural resources also reflects changing social concerns and outlooks. While the economy forged strongly ahead in the decades of the 1940s, 1950s, and 1960s, emphasis in the United States centered on institution-building, economic analysis, and increased equity. In at least the last decade other concerns arose. Trends evident in our growth suggested a too-rapid use of resources and a less-than-adequate concern for the environmental effects of that use. Yet we still do not possess an understanding of the interrelationship of the economy, the environment, and energy—and this is clearly central to materials policy.

Economic issues come into sharper focus through an example, namely, how the government should concern itself with materials industries. If industries are classified into three categories—(1) those that have been lost, such as zinc and leather; (2) those that may have difficulty in the future, such as steel; and (3) those that are moving ahead, such as electronics—it becomes clear that at least two policies are in order. First, we need policies that promote the successful industries so they remain at the forefront of trade. Second, for those industries in what might be called an endangered category, policy is needed as to whether, when, and how such industries should be kept operating in the United States. Such policy must relate also to foreign policy. Do we want an aluminum industry in the United States or do we want to help the lesser developed countries that supply bauxite? Is it in the interest of the United States to import at lower cost or to produce at home at higher prices? Should the United States have a policy for reviewing industries with the intention of deciding when an industry is in difficulty and whether it should be supported or allowed to drift out of the country? At the present time there is no mechanism even to consider such questions. At the same time, opinions differ whether this calls for a strong central materials policy or whether a pluralist approach would be more effective. At the least, however, these policy issues require further discussion and examination.

Energy

UMBERTO COLOMBO

Introduction

Over the long term, energy availability does not appear to be the limiting factor of development. Coal, with new technologies for its extraction and utilization, highly efficient nuclear fission technologies, the fusion of light nuclei, solar energy, and geothermal and other renewable energy sources offer not only practically unlimited quantities of power, but also a very broad and diversified range of possibilities from which very different scenarios can be imagined. A more efficient use of energy and a lifestyle that uses less energy to obtain greater return are certainly possible and offer another dimension to the long-term solution of the energy problem.

Eventually, factors other than the physical exhaustion of resources are likely to limit the development of energy sources: climatic and environmental constraints as well as technological, institutional, economic, social and political considerations will shape the energy policies of the future. No single long-range solution of universal acceptability appears to be technically feasible. Several possible solutions must be analyzed and their likely impact on the planet must be well assessed.

The main problem concerns the short and medium term, and it has two aspects: How can we "manage the transition" from the present age based on rapidly disappearing oil and gas reserves to other forms of energy? And how can we prepare a foundation for the future without closing today options that could become essential tomorrow?

It is my belief that an energy strategy for industrialized countries should be based upon three fundamental elements: (1) energy conservation over the whole cycle of production, including transformation and distribution processes as well as final use, with particular regard to the energy value of materials; (2) diversification of energy sources and, for imported energy, geopolitical differentiation of supply areas; and (3) careful recourse to nuclear fission, to cover those energy needs that cannot be satisfied otherwise.

As we shall see, science and technology, free of further constraints,

Umberto Colombo is Chairman of the Italian Atomic Energy Commission and former President of the European Industrial Research Management Association.

0077-8923/79/0334-0097$01.75/2 © 1979, NYAS

have an important role to play, both in helping us to face these problems and in solving them.

Energy Sources—Old and New

In the last ten years, the rate at which new oil reserves have been discovered has been lower than the rate of production. Even if some optimistic projections of new oil deposits in Mexico, in China and elsewhere were confirmed, this would not have any effect on the short term, and it would prolong for only a few years the availability of this source.

Technical, economical and environmental difficulties have limited the efficiency and applicability of secondary and tertiary recovery and the possibility of obtaining oil from low-grade deposits such as oil shales and tar sands. Past experience has shown a high rigidity in the oil market, and a gap between demand and potential supply can be predicted within fifteen years or less. The 1973 crisis and recent events in the Middle East have evidenced the highly political and complex nature of the oil supplies question. Although oil will continue to have a predominant share of the energy sources in the short and medium term, finding alternative solutions is a necessity, especially for those countries that have no or little oil reserves, like Western Europe or Japan. But the lesson we should have learned by now is more than just to move away from oil: leaning too much on one single source of energy for the bulk of our needs is a dangerous and unreliable practice.

The uneven distribution of hydrocarbons, in conjunction with the changing political scene on the world level and with the expectation that oil supplies (in the form of natural petroleum) will inevitably be exhausted in a matter of decades, has created conditions in which the oil market has become a seller's market whose laws are essentially dictated by the OPEC cartel, on whose wisdom we now must rely. In order to alter this situation, diversification of energy sources is an imperative requirement if we want to move gradually and without disruptive traumas toward an era of "definitive" energy sources. The best strategy for harmonious economic development must guarantee sufficient flexibility in the use of different primary energy resources.

Among alternative energy sources that can be exploited in the short and medium term, natural gas could in principle have an important role because of the relatively high consistency of its reserves; however, the possibilities for natural gas to assume a major share in the energy panorama are limited by difficulties of technology, economy, and safety that are connected with its long-range transportation and storage. Research and innovation can help in coping with these problems.

Coal could be adequate for several centuries, but its extraction and use pose a number of very difficult environmental, technical, and financial problems. As a consequence, the increase in the use of coal in these last years has been much slower than anticipated. Its conversion to gas and/or liquid fuel provides greater flexibility, but penetration of the related technologies into the market is not imminent on a large scale.

Nuclear energy has the potential of supplying an important contribution, but social acceptability poses a major obstacle to more widespread use. It is unfortunate that the energy debate in several countries has concentrated on the nuclear issue, thus dividing people into "pro-nuclears" and "anti-nuclears." This debate has stigmatized nuclear power, which—on account of its high degree of centralization, its capital-intensiveness, and the peculiar nature of its risks—has become the symbol of the rebellion of populations against the assault of technology on man and environment, and of the reaction of the man in the street against the arrogance of technocrats.

There are problems not totally solved as yet. They concern the nuclear fuel cycle, down to the disposal of radioactive waste, and, more generally, the safety considerations of nuclear plant operation, which must take into account "human" risk, which always goes together with "technical" risk. A resort to this energy source, however, is necessary, particularly in industrialized countries, to provide a viable solution to the energy problem. The need that our society has for nuclear energy must, however, be clearly explained, so that it may be understood by the public at large. Nuclear power, in other words, should be considered part of the "normal" complexity of our technological society.

Because of the inadequacy of market-economy mechanisms, solar energy and other possible renewable resources are likely to have a slow penetration into the market, even for those applications that are already reaching industrial production. They require structural changes that cannot be implemented easily. In the near future, these new sources will probably contribute at most a small portion of the total energy supply.

Governments will have to play an essential role of extreme importance in accelerating the development of these technologies and in creating or stimulating a market for their expansion.

The Role of Energy Conservation

Is then energy conservation the solution? The increased cost of energy, the incipient effects of market saturation, and increased marginal costs for the installation of new power plants or the extension of distribution networks, all have been instrumental in encouraging more efficient use of energy, a reduction of waste, and a decrease of energy-intensiveness in the production of goods and the provision of services. Such a tendency is apparent in a greater or smaller degree in all industrialized societies. Although market pulls and economic forces are active in this direction, one cannot rely on them alone to implement the conservation measures that are needed. These forces respond to short- and medium-term signals, and do not promote the more basic changes and the planning or large investments that may be required. Recovery of heat, elimination of heat losses, new heating systems and, more generally, better energy management practices can bring consistent energy savings in most sectors.

Further savings can be achieved by modifying or improving existing processes or plants, by matching characteristics of energy sources with users' requirements (such as temperature), or by changing the products and services through new investments. All this also means investing in energy today to save more energy tomorrow.

Because the evaluation of return on investment is the main criterion in deciding on new opportunities or initiatives, market forces alone, or short-term economic considerations, cannot induce the transition, as already mentioned. On the other hand, an energy-conservation policy of this sort cannot be too rigid or implemented too quickly, not only in view of the investment required, but also on account of the negative effects it could have in the short term on production and employment.

This situation might be somewhat different in the less developed countries, particularly those rapidly becoming industrialized and urbanized. Generally, in countries such as India, Brazil, Argentina, Mexico and others of similar economic importance, because of industrial consumption and the rising living standard, the growth rate of energy consumption is currently higher—by 50 percent or even more—than that of their gross national products and it generally implies doubling times of 10 years or less. However, these countries would still have the possibility of basing their development on less energy-intensive processes. Rather than continuing uncritically on the basis of past history and models established by the industrialized countries, use should be made of their past experience, both negative and positive. For both developing and developed countries a global strategy is needed that takes into account all the relations between energy sources and economic, social and political problems.

Looking for a Strategy

A rational energy strategy would require concentration of efforts on one of a few well-defined options in order to reach a maximum of results in the shortest time. It seems very unlikely that present societies are able today to formulate and to accept choices that could become in fact irreversible, such as the "all-nuclear" option accompanied by a large increase of electricity and the introduction of other energy carriers such as hydrogen or methanol, or the opposite option based on the hope of an early development of renewable energy sources and of a quick increase in the efficiency of energy utilization.

It seems therefore inevitable that we will pursue efforts in several different directions, postponing choices to a later time. This broadfront strategy would certainly be facilitated by a high degree of international integration. To reduce multiplication of efforts and to make such a strategy compatible with the available economic and human resources, not all options should be followed by all countries at the same time. Results of investigations, demonstrations, and practical experience with deployment of a source should be widely shared with other countries, so that decisions on implementation elsewhere could be taken on a firm basis. The practicality of such strict international cooperation, although desirable, is highly questionable.

It seems, however, necessary at the moment not to neglect any possible source of energy and to keep open any viable option. Hard and soft technologies are not incompatible but rather complementary. As stated by Brian Flowers, nuclear energy should be seen as a useful contribution to the conservation of resources. Even minor energy sources, or conservation measures, may allow us to "buy time" so that we can come up with the technological innovations that will in turn make possible the necessary socioeconomic evolution.

The Role of Energy Research

This takes us to the double role of research in this context: to help manage the transition period, and to prepare new options. The difficulties of this endeavour derive from the long lead time that is necessary for research to obtain useful results, and that adds to the time needed for the demonstration and generation of innovations in the marketplace. They also derive from the limited amount of capital and human resources available. To break this impasse without undue limitations on future choices, two conditions are necessary: one is the clear definition of the objectives in each field and of the time scale involved in achieving them; the second is broad cooperation and, whenever possible, a true integration of efforts at the international level.

For instance, in the field of nuclear energy, the shorter-term research should be focused on the safety (and security) aspects of the proven reactor types and of the different phases of the fuel cycle. Although safety has been considered a high-priority area by the nuclear industry, which has reached an unsurpassed safety record vis-à-vis most other industry sectors, the risks connected with nuclear energy have very special characteristics: one can suppose an accident with very serious consequences but with an extremely low probability. The evaluation of highly improbable situations with large negative effects is outside common experience; it is therefore natural that the public demands much more detailed and convincing evidence regarding nuclear energy than is required for other sources whose safety and environmental features are perhaps worse but closer to everyday experience. A number of research programs are under way in this area, and there is a certain degree of international cooperation; more should be done, however, and objective information in terms easily understood by a lay audience should be widely circulated to gain greater public acceptance for nuclear energy.

Another objective of short- and medium-term research in the nuclear field should be an improved utilization of nuclear fuel in the present reactors; in this field, very different views held by various countries on the issues of proliferation of nuclear weapons, and on measures to prevent it, currently do not allow a truly international coordination of efforts. Hopefully, the conclusions of the International Nuclear Fuel Cycle Evaluation Program, due one year from now, will provide some new insight into this highly debated issue and allow new constructive initiatives. Better fuel utilization would allow extra flexibility in nuclear strategy and postpone more radical solutions until technologies are more mature.

If the nuclear option has to be kept for the long term, there seems to be no alternative to breeder reactors; here again, if political conditions allowed a greater coordination of research in this field, it might be wise to explore more than just the sodium-cooled fast breeder as is being done today in several independent programs in various countries.

With solar energy, short- and medium-term utilization will probably be concentrated on decentralized small-scale applications in the areas of water heating, space heating (and, perhaps, cooling), agricultural needs, and small electricity generators in isolated or remote areas. Acceptance of distributed collection of solar energy will depend on the extent to which users are individually relieved of the problems of financing installation and servicing. The main concerns in this field are the optimization of systems and designs, engineering improvements, and large-scale manufacturing of equipment to reduce costs. Passive users of solar energy can be more important in the medium term than active ones, and they require research and development in the design and technology of construction of buildings.

In view of longer-term and larger-scale applications of solar energy, many options, both hard and soft, are possible, including thermal conversion, photovoltaic systems, and biomass processes, each with a large variety of alternatives. It is important to clarify the advantages and disadvantages of each, and to select the appropriate solution at the right time, which may vary according to the specific problem to be solved and to different local conditions. In this field, however, there seems still to be space for innovative and creative thinking, and although the hope for a breakthrough that would radically alter the situation is smaller today than it was only a couple of years ago, the possibilities are certainly more open than in most other fields.

Research and development (R&D) can serve to identify and reduce waste and inefficiency. But we can obtain concrete results only if scientific and technological aspects are integrated into the framework of a more global policy that favors the restructuring of the economy and industry, and if we are more frugal with energy than we have been hitherto.

Even the most optimistic supporters of nuclear fusion consider it a long-term option, but the diversity of this approach, and its promise of an unlimited energy supply, justify the large amount of research that is carried out in the world in this area. There has been an oscillation in recent years in the opinions regarding the necessity of moving from basic physics studies into technology: on the one hand, technological difficulties may be the limiting factor which should be identified as soon as possible; on the other hand, an early emphasis on technological development may bring about a concentration on just one concept that has not yet proven to be necessarily better than others for the long term. The solution may come again from international cooperation, which has traditionally been very strong in this field. The recent initiative for a worldwide concentration of efforts on one prototype Tokamak machine of the next generation (the INTOR project) could free investments and resources to explore other possible solutions at a lower level of effort.

The critical period ahead of us concerns the medium term, a time when society does not yet have the definitive technologies available, and energy intensiveness in the economy is still very high; the scarcity of oil and the consequent rise in its price could determine a financial crisis of very large proportions. Such a crisis would involve as one of its effects the necessity of concentrating efforts in research and development in energy as well as in other fields, and of selecting only a very few of the available options. If research can do little to prevent this crisis (which could perhaps be avoided by a high degree of international cooperation and interconnection in other fields) it should at least provide sufficient preparation to enable options to be chosen in the most rational way.

THE LONG-TERM ALTERNATIVES

Haefele and Sassin have provided a useful quantitative framework for the assessment of long-term energy supply options, describing a solution that involves the establishment of an energy endowment based essentially on coal, solar and nuclear-fission energy. This scenario, placed in the year 2030, seems to be compatible with fairly consistent economic growth at the world level over the coming decades. It allows for a per capita energy consumption in the less developed countries, over a span of 70 to 100 years, that is equivalent to the present consumption of Europe and Japan. The Haefele-Sassin model assumes essentially a continuation of the basic trends of the past model of development of industrial countries. It proposes mechanisms such as power production located in remote sites, nuclear and hard solar-coal hybridization that in fact do not depart from the trends toward the centralization of production and the structural rigidity deriving therefrom. The necessary amount of capital is estimated on the order of $40,000 billion (constant 1975 United States dollar value), $8000 billion of which would be utilized for the less developed areas. The financial, technical and organizational difficulties to support a program of these dimensions would, however, require a high degree of world order and a large degree of control and regulation. Such a new global management system to cope with world energy problems (capital investment, research and development, security and safety) would have very serious implications for the sociopolitical structure of the world. The preoccupation that our society might be induced to drift gradually toward authoritarianism has promoted Dr. O. Bernardini and myself to work on a scenario characterized by a much lower energy requirement vis-a-vis that of Haefele and Sassin.

Fifty years from now, energy needs could be much more contained than those foreseen by an extrapolated development, and not far removed from the present per capita average, but rather better distributed. This kind of pathway would guarantee for everybody a satisfactory life standard with no further heavy exploitation of resources.

Our industrialized society is becoming saturated with material goods, but demand grows in the direction of information-rich goods, for which the purely material and energy content become less and less important, and of nonmaterial goods such as education, health, a clean environment, social security, and the right to take part in decision-making.

Among the more topical and promising issues there is the development of low-energy-intensive activities, such as characterize information technologies and biotechnologies. Information technologies require a small investment of energy, but have an extraordinary capacity to influence the efficiency of economic activities by reducing energy consumption and wastes. Energy and information are strictly interconnected, and stressing the one with respect to the other is progress towards high thermodynamic yields.

Biotechnologies, still at an early stage of development today, are already used in agriculture where they increasingly substitute for chemicals and mechanization. Furthermore, biotechnologies can be directly employed in energy production, for example by extracting alcohols and hydrocarbons from vegetable matter.

These highly innovative technologies can lead to an active conservation of natural resources. Furthermore, their development could prevent some large-scale negative effects and allow the attainment of a valid decentralization. The issue of centralization versus decentralization implies balancing benefits: technological standardization against self-involvement of populations, and homogeneous criteria of safety and security control against the opportunity for the man on the street to assess the risks and benefits of different energy options. However, decentralization may lead to a better use of most resources (energy, materials and territory) and can also contribute to the solution of the serious unemployment problem that our society faces today.

CONCLUSIONS

R&D strategies in the energy field have an unusual nature. The problems are generally different from one country to another, due to large variations in resource availability, industrial structure, economics, climate, environment, culture, institutions, and so forth. On the other hand, as we have seen, more extensive international cooperation is needed to share burdens and to keep viable options available.

Although international institutions to coordinate energy research do exist (such as the Organization for Economic Cooperation and Development and the agencies of the United Nations), their role has been largely marginal in comparison with national R&D policies. Each country has preferred to maintain complete control over its programs, and mutual trust in sharing the results of efforts carried out elsewhere has not been strong enough to justify abandoning independent national efforts.

As a result, a large duplication of work is apparent, while many areas of potential interest remain unexplored or not sufficiently investigated. Unfortunately, the solution to this dilemma is not simple. It cannot simply be brought about by the establishment of another international agency; it requires a substantial change in the international climate.

Energy R&D cannot be seen in isolation: energy production and use interacts very strongly with the availability, production, and use of other resources such as minerals, water, air, food and land. Therefore, a policy of energy R&D must be closely integrated with policies concerning these other fields.

For the medium term (until around 2010) one can envisage: (1) using energy sources that are now available, making recourse to innovative technologies, (2) developing a number of new energy options, (3) starting to use new energy carriers, and, (4) introducing substantially new processes for the production of some goods and services. For oil, this means tapping its potentially large offshore deep-sea deposits, and possibly developing *in situ* processing of oil shales. For coal, it may include automation of mining, gasification and liquefaction processes, *in situ* conversion, or advanced transportation methods. In the field of nuclear energy it may involve advanced converters and breeder reactors, possibly new methods of uranium enrichment, and development of satisfactory solutions for final waste disposal. For other energy sources it means pursuing large-scale solar applications, including photovolatic and biomass systems; extended use of geothermal energy, in particular, hot dry rocks and low-enthalpy deposits; development of other decentralized energy sources, especially for applications in rural and remote areas; and general studies on prospective energy systems.

For the long term, options are still rather open and vague; nevertheless, it is necessary now to put a substantial amount of effort into R&D to prepare alternatives for the future. A choice will have to be made later, and it will depend not only on external conditions, but on the results of R&D itself. It should not be forgotten that all research contains an element of unpredictability, and that it is necessary to update a program continuously to keep account of the results.

The question about which model of society we should try to develop might seem like idle talk. However, some sort of a long-term model can help us with our short- or medium-term decisions.

The needs—economic growth and conservation of energy and

materials, reduction of disruptions, respect for cultural values and for local conditions—are numerous and interrelated. They refer to situations characterized by profound differences which certainly shall be considered. In any case, dead-end roads or rigid and constraining paths should be avoided.

It is indispensable, therefore, to have a whole set of alternative options at hand, so that different technologies shall be available to satisfy the evolving demand, in a sort of "technological pluralism." For example, it is by now clear that the "optimal" energy source does not exist. When a society is too much conditioned by a "unique" choice considered optimal, it easily becomes too rigid and vulnerable. The example of the conditioning choice of petroleum as the hegemonic and prevailing energy source is enlightening. The coexistence of soft and hard technologies, and of both centralized and decentralized systems, makes a more diverse society possible and allows a balanced evolution.

This pattern of development in advanced countries also calls for fewer resources, including capital, for its achievement, and hence leaves greater room for the growth still required by the Third World, which would otherwise prove to be highly difficult and unrealistic.

Overview of Policy Issues:
Panel Report

Panel Chair: EDWARD E. DAVID, JR.
Panel Members: Dominique Akl, Arthur M. Bueche,
Philip Handler, Robert I. Hanfling, Richard F. Hill, Gösta
Lagermalm, Henry R. Nau, Dorothy Nelkin, J. E. Penick,
Umberto Ratti, and Benjamin S. P. Shen
Rapporteur: Ted Dintersmith

TRANSITION ERA

DURING THE NEXT FIFTY YEARS, the world's economy will be
making will be making a transition away from today's conventional
sources of energy. Exactly how that transition will evolve and where it
will lead hinges upon a multitude of factors, one of the most important
being science and technology policy. A discussion, then, of science and
technology policy with respect to energy issues cannot be conducted for
1979 alone; it must lie in the context of more than a half-century of
societal evolution.

The images of our energy future are unavoidably vague. Technologies
being developed today—photovoltaics, biomass, solar, fusion, the
breeder reactor, solar thermal, and passive design—may play substantial
roles in meeting the world's future energy needs. But the growth and
development of these technologies or of yet undiscovered technologies
are highly uncertain, depending as they do on unknown economics, unpredictable social and political factors, and international politics. Science
and technology policy has no choice but to recognize this complex,
dynamic, and uncertain environment.

One inescapable certainty that policy-makers must address is that the
options of the distant future are precisely that. Any hope of utilizing fusion, photovoltaics, the breeder, biomass, solar, or solar thermal energy
on a large-scale economically feasible basis in the next two decades is
based on a thin thread of optimism, as is the hope for massive reductions
in energy demand through end-use conservation. The world's energy pic-

*Edward E. David, Jr. is President of Exxon Research and Engineering Company
and Vice-President, Science and Technology of the Exxon Corporation. He is a
former Science Advisor to the President.*

0077-8923/79/0334-0108$01.75/2 © 1979, NYAS

ture for the rest of the Twentieth Century will be dominated by petroleum, coal, and nuclear fission energy sources. Science and technology policy should examine how these conventional sources can be most effectively utilized while further information about alternatives is gained through research, development, and pioneering deployments.

It is clear, then, that science and technology policy must take explicit account of the dynamic aspects of the world's transition from the current energy mix to a much different future. The fact that budgets for research and development (R&D) are not unlimited means that trade-offs between long-term and short-term options must be made. A policy that blindly pursues one course, at the expense of the other, will be vulnerable to failure.

The interrelationship between science and technology policy and the resource-management decisions of those nations holding substantial reserves of conventional sources (such as the OPEC nations) further complicates the transition policy. From the perspectives of the nations of the Organization for Economic Cooperation and Development (OECD), science and technology policy could be more clearly charted if the future price and supply decision of OPEC nations were known well into the future. However, from the OPEC nations' perspective, their petroleum production schedule and pricing policy could be best planned if the future of science and technology and economic policy of the OECD nations were known. A policy that emphasizes long-term alternatives would indicate to OPEC that it would be advantageous to them to charge very high prices in the immediate future. A science and technology policy that focused on the short term would signal OPEC that a delay in production would be relatively profitable.

The interrelationship between science and technology policy and OPEC is just one of many factors that underscore the complexity of science and technology decisions. The clear implication of this problem structure is that, with respect to energy concerns, science and technology policy should be conducted on an incremental basis. At frequent intervals, recently acquired information produced by research and development, as well as new information about the international energy environment, should be used to modify and reorient the priorities of science and technology policy.

Of equal importance will be information about transportation options, the needs of industrial production technologies, and health effects. Factors such as these, which are directly and indirectly related to energy technologies, will change in uncertain ways; information that may be revealed in the future will inevitably influence science and technology policy. Hence, a policy that is conducted on an incremental basis is judicious. Yet the time scales for technology development and deployment are much greater than those for the aforementioned transients. So the science and technology strategy must be relatively insensitive to such changes. A robust strategy is the most critical need for the future.

In formulating a robust strategy for R&D investments in various technologies, science and technology policy-makers should adopt a systems approach. That is, the cost and expected benefits of R&D for each technology should not be weighed in isolation. A broad framework, encompassing all options should guide decisions. Using a systems approach in formulating science and technology policy in the area of energy R&D has numerous advantages, including the following: (1) priorities can be set among the candidate technologies; (2) a diverse portfolio of high-risk and low-risk R&D investments can be constructed; (3) a balance between longer-term and shorter-term technologies can be struck; (4) the entire spectrum of technologies, not just the "glamorous" technologies, are taken under consideration; (5) R&D investments among different technologies can be coordinated; (6) basic research that may benefit several technologies can be emphasized; (7) information gained from R&D in a particular area may be used to guide the overall R&D strategy; and (8) the risks of having or not having each technology can be estimated on a more rational basis.

While the systems approach to R&D decision-making is valuable, organizational realities—in both the public and private sector—may impede this approach. If the activities of the Federal government's agencies continue to conflict, inefficiencies in R&D investments will result. In the private sector, a particular corporation may approach its R&D decisions in a systematic way, but R&D conducted by the private sector as a whole may not be achieving its full potential.

It may well be worthwhile for science and technology policy-makers to examine current R&D practices in government and private industry, looking for R&D investments that lead to inefficient energy-supply usage systems. Policies that can facilitate a more diverse and systems-oriented approach to R&D, within both the public and private sector, should be identified and examined.

A systems approach to science and technology policy would also allow policy decisions to be integrated with the surrounding social system. Science and technology policy should both reflect and guide society's preferences. For example, broad choices between decentralized energy systems (emphasizing soft solar technologies) and centralized systems (perhaps involving large energy parks or islands) cannot be made on energy criteria alone.

Similarly, the role of conservation in future energy use is not simply a function of the cost of energy. Judgments made by people with respect to the life-styles they desire will, to a large extent, determine the extent to which conservation can curb energy needs.

The challenge that lies ahead for science and technology policy in the area of energy is immense. Decisions made today have implications for the next fifty years. During this half-century, the world will be shifting from today's conventional sources of energy to alternatives that are still being researched. The role of energy in society's functioning increases the challenge. Energy is not a discretionary commodity; its form and the extent to which it is available are interwoven in the fabric of industrial and domestic life-styles. Consequently, the issues facing science and technology policy-makers are multiple and complex; vision and creativity are essential.

THE ROLE OF GOVERNMENT AND THE FREE MARKET

An analysis of science and technology policy should examine exactly what levers can be pulled by policy-makers. In the United States, controls lie largely in the hands of the Federal government. Since innovative new technologies will be developed principally by private corporations, Federal science and technology policy-makers must strive to produce a climate conducive to such private commercialization.

In the United States, major technological breakthroughs have generally been accomplished without large government subsidies. The development of the automobile, the efficient conversion of electricity from fossil fuel, and the advance of computers are representative examples of technological innovation pioneered in the private sector. A strong free-enterprise system that rewards entrepreneurs who combine technological possibilities with market demand is a very powerful innovative force.

In a free-enterprise system, price serves a valuable function. A high price for a good signals entrepreneurs that the development of alternatives will be profitable. Conversely, low prices discourage the development of alternatives.

In the United States, science and technology policy has had an extra hurdle placed in its path by energy-pricing policy. For thirty-five years, energy has had a regulated price. Most notably, in recent years, the prices of oil and natural gas have been held at artificially low levels by Federal policy. This policy has two unfortunate consequences: more energy is used, because its price is subsidized; and, entrepreneurs hoping to develop alternatives are hindered by the artificially low price of existing supplies.

Recent policy changes to deregulate the price of oil and natural gas will boost the rate at which alternatives are commercialized. As the price of oil and natural gas increases, alternatives such as coal gasification, shale oil, and solar thermal energy become increasingly attractive.

Unfortunately, government policy seems unable to correct past discrepancies. Since oil and gas prices were held at artificially low levels in the past, thus hindering the development of alternatives, the current temptation is to subsidize the price of the alternatives. Such subsidization may be attractive in the short run, but inevitably this distorts the long-run situation. The tasks facing science and technology policy are severe enough without the added complications of artificially determined prices for the competing technologies.

Particular policies to be avoided are those that restrict the choices available to consumers. Steps that mandate certain actions—what types of automobiles can be bought, how new houses must be constructed, what types of fuel can be used by industry—can be counterproductive. The free market system is the best mechanism available for allowing consumers to make efficient decisions based on their own needs. As long as prices reflect the full cost of production, including environmental costs, the free market system can efficiently administer energy "policy."

The private sector's ability to innovate has been called into question recently, particularly with respect to projects whose payoffs are highly uncertain or at least unlikely to be realized in the short term. This view is used to justify much of the research and development work done in the public sector. That is, according to this perspective, the government should carry out R&D on any project that has benefits that will not be realized in the near term or where the risks are high.

Considerable evidence exists, however, that indicates that private industry will and *is* doing much risky research and development work that is not expected to yield benefits in the short term. Moreover, when industry itself performs the task of research and development, commercialization is expedited. If, however, government assumes an active role in carrying out R&D in particular areas, the incentive for the private sector to do similar R&D is lost.

Whereas private corporations are capable of performing a large share of the R&D needed to bring energy technologies to commercialization, the Federal government can take decisive steps to speed this innovative process. In particular, private entrepreneurs depend heavily on the existence of a pool of basic science knowledge. Little basic research is performed by the private sector; most is conducted by our nation's universities. The Federal government should ensure that universities receive healthy support for basic research programs. Furthermore, this support should afford recipients the flexibility and freedom to pursue their goals regardless of where their channels of discovery may appear to lead. In view of this, the Department of Energy's recent move to strengthen its basic research is encouraging.

When the private sector is developing large-scale projects, some government risk-sharing is needed. No single corporation is likely to assume all responsibility for front-end financing of a billion dollar demonstration plant if its commercial or technical success is questionable. Without Federal assistance at this phase of the research, development, and commercialization process, the feasibility of a technology on a commercial scale may never be established and the technological innovation process is not complete until the technology has been implemented on a commercial scale.

A specific example of a technology that the private sector, with Federal assistance, can and should be demonstrating is coal gasification. Other promising areas that might benefit from this industry/government approach include those of coal liquids, cleaned coal solids, and shale oil.

The importance of commercializing some of these promising technologies cannot be overemphasized. If the United States, for example, had an incremental 1 to 2 million barrels per day of synthetic fuels, OPEC's grip on the United States economy would be loosened. Science and technology policy, then, clearly should emphasize rapid commercialization of promising energy technologies, and this requires substantial financial commitments in the demonstration phase.

Thus, science and technology policy should focus on steps to create a climate that encourages innovation in the private sector. Beyond support of basic research, key variables under the government's control include wage and price policy, regulatory policy, and tax policy. First priority should be given to increasing the stability of Federal policy in these areas. Entrepreneurs considering the undertaking of risky innovation in energy technology will be discouraged by vacillating Federal policy.

As science and technology policy gradually moves Federal activities out of the development area, and places added emphasis on basic research, the private sector will assume added R&D responsibilities. However, an increased R&D effort will require additional investment in the commercialization phase. The Federal government, through its tax policy, should take steps to ensure that private corporations can form the capital needed to invest in the innovative process.

COOPERATION

If science and technology policy is to deal effectively with energy concerns, cooperation between various groups will be essential. Cooperation between citizens and decision-makers will be imperative if energy decisions are to be made credible. Science and technology policy should actively promote cooperative activities involving people from government, industry, and universities. Finally, international cooperation can play a major role in facilitating the world's energy transition.

The social sciences can contribute immensely to science and technology policy in the area of energy. The interactions among experts, decision-makers, and the general public should be investigated by social scientists. Particular attention should be focused on how the decision-making process reflects the availability of information, communication bridges, and overall social preferences. Experiments in communication should be conducted, and their effects evaluated.

It seems clear that the public's understanding of the technologies, as well as the technologist's understanding of the public's preferences, are all too vague at this point. Considerable attention needs to be paid to ways in which the interaction between the public and the technologists can be strengthened. In particular, the effectiveness of our Federal government in acting as a broker for these two groups, and how this linkage can be improved, warrants careful research.

The creation of research teams constituted of people from the public, private, and academic sectors offers tremendous potential for progress in the area of energy-related science and technology. Drawing together members of each group—with their different backgrounds, perspectives, and skills—in a cooperative environment would likely produce exciting results. The benefits of this approach would probably extend beyond the synergism of a multifaceted research team; uniting different groups in one cooperative environment would demonstrate that different national groups could team together to penetrate difficult and crucial problems.

Cooperation on the international front also holds considerable promise for advances in energy-related science and technology. Whereas the current extent of cooperation is high, particularly in the private sector, the saturation point has by no means been reached. Research in areas of complex, large-scale systems, such as fusion, are conducive to East-West cooperation. North-South cooperation can play a vital role in the development of decentralized/renewable energy sources. Federal policy in this respect should address questions about proprietary rights and the extent of government oversight and "red tape" required for international joint ventures.

Conclusion

The world is currently in the initial phase of a 50- to 70-year transition from conventional sources of energy to one of so-called "sun/fuse" technologies. The evolution of this transition and its ultimate destination will be profoundly affected by our science and technology policy. In rising to this challenge, science and technology policy must adopt an incremental approach to decision-making, using to the maximal extent possible whatever new information becomes available about particular technologies, international energy developments, and society's preferences. Yet the long time frame for technological innovation calls for a robust, stable strategy. Science and technology policy will be much more effective if it views competing technologies in a systems framework; decisions about particular technologies should be formulated in the context of overall energy technology strategy.

The free-market system is an efficient vehicle for bringing innovative technologies to commercialization. Federal responsibility should focus on basic research, sharing the risk of costly commercializing and risky technologies where necessary and using tax policy to ensure that private corporations have the investment funds needed to spearhead the innovation process.

Existing evidence indicates that the funding devoted to R&D by the private sector is adequate. Other aspects of the developmental climate—government-controlled energy prices, vacillating regulatory policy, uncertainty about wage and price policy, and mandatory requirements involving the types of energy that can be used—appear to be impeding the innovation process. Science and technology policy should examine the overall climate for innovation with respect to energy technologies, not just the availability of R&D funds.

Finally, the benefits of cooperation have not been adequately tapped. Science and technology policy-makers should facilitate interactions between technologists and the general public to improve mutual understanding. A climate in which these groups view each other as insensitive adversaries cannot be productive. The scale and complexity of energy technology problems is imposing; if science and technology policy can take the lead in creating research teams consisting of people from government, industry, and universities, these problems will be resolved much more quickly. The benefits of international cooperation are also extremely attractive. A science and technology policy that emphasizes cooperation will achieve not only greater progress, but also improved understanding on the part of each group of the others' views.

The Environment

DAVID SWAN

So MANY PROBLEMS and issues have been raised in the field of environmental policy that identifying significant problems in the area and mechanisms for the application of technology policy poses an exceptional task. To make this complex topic more manageable, let us proceed by considering United States environmental policy and then examining the major potential opportunities for interaction with the science and technology community. I have selected the United States as a focus because our country has adopted by far the most comprehensive set of environmental policies and therefore necessarily can serve as a standard against which other policies may be compared.

In the late 1960s, through Congressional initiative, the United States embarked on a massive Federally directed program to improve environmental quality. Each session of Congress since 1970 has produced major legislation which, by 1976, resulted in an intricate legal structure designed to ensure attainment and maintenance of a superior environment. A partial list of the significant laws now in place includes: the Clean Air Act of 1970, amended in 1977; the Clean Water Act passed in 1972, amended in 1977; the Safe Drinking Water Act; the Toxic Substances Control Act; the Resources Conservation and Recovery Act; the National Environmental Policy Act; and the Occupational Safety and Health Act. These are the major laws; in addition, there are a host of other specific acts dealing with particular problems such as the Surface Mining and Control Reclamation Act and the Mine Health and Safety Act.

One major consequence of this massive Congressional initiative was to set up a major new bureaucracy, the Environmental Protection Agency (EPA), as well as to add significantly to the regulatory authority of other groups, including the Department of Labor, the Department of the Interior and the Department of Energy.

In all cases the laws delegated preparation of regulations and enforcement through a complex state/Federal system accompanied by a strong effort to involve the public at all stages in developing and enforcing

David Swan is Vice-President, Environmental Issues, of the Kennecott Copper Corporation.

0077-8923/79/0334-0116$01.75/2 © 1979, NYAS

regulations. Despite the complexity introduced by these multiple procedures, tight attainment timetables were also specified in most of the
laws, and extreme sanctions were imposed for violations, including
criminal penalties, stiff fines and possible jail sentences for offenders.

The basic policy embodied in the framework of the laws is to achieve
environmental cleanup by reducing or eliminating emissions of gases and
liquids and by providing for permanent storage of wastes in an inert
form. In addition, the Toxic Substances Control Act provided for track-
ing toxic substances during their entire life from initial production
through eventual disposal.

Congress provided that public sector expenditures, primarily those for
municipal water and sewer systems, be funded by Federal grant programs as well as local taxation, and adopted a "polluter pays" policy for
the private sector, which requires the "source" of the pollutant to pay for
investment and operating costs for the equipment necessary to curb emissions and effluents. Limited tax relief was included in several laws,
primarily by permitting favorable financing terms for required capital investment. Although enforcement is delegated to state bodies, there is
provision for direct enforcement by governmental agencies in the event
that the state or local authorities would not or could not enforce the
regulations.

Obviously, such a massive effort is costly. For example, the Council
on Environmental Quality estimates that cumulative expenditure in the
decade 1977 through 1986 will amount to $645 billion, expressed in 1977
dollars. In actual dollars this is likely to be well over a trillion. Perhaps
more importantly, even in constant dollars, the Council sees investment
and operating costs accelerating to an annual rate of $85 billion in 1986.

In the rush to provide a comprehensive environmental package a
number of major issues were dealt with rather superficially. Clearly,
there could be no real knowledge of costs or benefits since both depended
on regulations whose preparation was delegated to the EPA or other
government agencies. Little or no recognition was given to the profile of
economic or other impacts as a function of geography or of industry sector, or for that matter of differences in climatology or other major environmental factors. Although environmental protection is clearly a
global concern, the United States approach paid little attention to the
potential or actual actions of other countries. Despite extremely tight
timetables mandated by the laws, no specific Congressional direction
was given regarding the priority of effort by the enforcing bureaucracies.
Finally, in setting up the enforcement agencies, which were fully preoccupied with organizing and establishing their new responsibility, little at-

tention was paid to interagency coordination among the various groups responsible for framing and enforcing regulations.

Despite these defects in the Congressional approach, the country has made significant progress in cleaning up major emission sources, particularly in the area of water, and it is clear from repeated surveys that the American public wants a cleaner environment.

What then are some of the major issues that must now be dealt with? Direct problems are fairly easy to identify, although difficult to solve. Perhaps the most important is to determine with greater certainty the health effects from long-term exposure to low concentrations of pollutants and classes of pollutants.

Prior to the early 1970s, most exposure standards were based on toxicological tests designed to elicit responses of human or animal populations to specific doses. In framing the standards mandated by Federal law, EPA relied on epidemiological approaches, previously not used for this purpose. Better and more objective criteria for establishing action program priorities are badly needed.

In the absence of clear Congressional direction, and given the very ambitious timetables and extensive opportunity for the involvement of the public, the legal system is increasingly being relied upon to enforce more rapid promulgation of regulations or to delay regulation or enforcement action. Clearly, institutional methods of coping with this problem must be evolved. Among the many suggestions for dealing with the problem are the creation of science courts, appointment of ombudsmen, and development of arbitration techniques. Congress itself has increasingly involved the National Academies as advisors with respect to health effects, a role that is not particularly relished by that august body, and on at least one occasion Senator Muskie wished aloud for a few one-armed scientists after hearing repeated testimony regarding the uncertainty of health-effects judgments. Our universities have, in general, steered clear of this contentious issue, although individual faculty members frequently serve on technical advisory panels and a few public health schools have embarked on ambitious research programs.

Industry has reacted by setting up research institutes such as the Electric Power Research Institute and the Chemical Industry Industrial Toxicology Institute.

Government agencies, primarily the National Institute of Health, National Institute of Environmental Sciences, EPA laboratories and the National Institute of Occupational Safety and Health have done the greatest amount of work, although a considerable amount of this has been questioned by scientific peers.

The second major technical issue deals with the source and fate of pollutants in the environment. Again, the principal actors are the government agencies and private industry.

A third area now receiving increased attention is the relationship among costs, risks and benefits. Development of acceptable methodology is badly needed and dissemination of current techniques has been spotty. Again, our academic institutions have been relatively inactive in dealing with these problems.

Development of new process technology for dealing with the capture and concentration of pollutants is an extremely active area primarily served by the industrial sector, particularly that portion which views it as a new business opportunity. Significant technical progress has been disappointingly slow and very little modern technology has been produced or applied, probably because the extremely tight timetables made the application of existing technical solutions to pollution problems almost mandatory. As new plants are constructed we can expect to see more innovative technical solutions adopted, particularly in the basic industries, and involving novel methods for concentrating waste streams. Again, the industrial sector is providing most of the technology.

In summary, the major technical problems associated with United States environmental policy are being attacked in a relatively conservative fashion and there is some evidence that industrial restructuring is taking place. Academic institutions have not significantly changed their role, particularly when compared with other professional institutions such as the legal profession. Professional societies have also been active in bringing together the regulators and regulated.

The indirect effects, however, of environmental policy are both more subtle and perhaps more far reaching. In the private sector the principal burden of environmental regulation falls on the manufacturing industries, with the greatest impact on the basic industries: steel, nonferrous metals, chemical, pulp and paper, petroleum and the electric utilities.

Present United States environmental policy makes it difficult and expensive to construct plants to process basic materials. Because raw-material-rich countries, particularly in the Third World, have not assigned high priority to environmental regulation, a basic restructuring of the natural resources industry may be under way, particularly in the metals sector. The consequences of this restructuring on national security and assured availability of mineral and energy resources can be extremely significant and comparable to the example of OPEC.

The effects of massive and detailed regulation on productivity and innovation are only now beginning to be perceived. Since in any advanced industrial society innovation is critical to improving productivity and thus improving the general standard of living, this may be the greatest impact of all. A major Presidential policy review is now under way on this subject and should be available shortly. Clearly, this should be a matter of great concern to all of us dealing with science and technology.

Many of these problems have been recognized in the Executive Branch and in Congress, and we have recently seen the formation of a Regulatory Council, Regulatory Analysis and Review Group, and the announcement of increased activity of Congressional oversight in the regulatory process.

In summary, our scientific and technical institutions clearly have been reactive to the problems and opportunities posed by the environmental movement, particularly when contrasted with other institutions in our society. While this is a somewhat critical assessment, there are obviously many opportunities where the science and technology community could take a more active role in the resolution of the difficult tradeoffs that must be made. One point appears clear: that the country has no intention of making radical shifts in environmental policy and, therefore, any changes must be evolutionary rather than revolutionary.

Overview of Policy Issues: Panel Report

Panel Chair: MERRIL EISENBUD
Panel Members: Laurence Berlowitz, Sidney Borowitz,
Dennis Dugan, Milton B. Hollander, Franklin A. Long,
Seymour L. Meisel, Charles Mosher, Richard B. Opsahl,
Lewis H. Sarett, and Paul Silverman
Rapporteur: Richard Langlois

AMERICA IS, IN MANY WAYS, a society regulating itself far beyond the point of diminishing returns. Regulation is stifling innovation and wasting human and material resources.

These are perhaps the central conclusions reached by the environment panel in its discussion following the presentation by David Swan. The environment—narrowly construed under the aspect of pollution abatement and control—has been the motivation for massive investments by governments and industry over the last decade. But it is not at all clear that society has reaped public health benefits from this investment that are commensurate with its cost, or, in some cases, that there have been any benefits at all.

The panel discussed a number of specific reform proposals and agreed on the important role for science and scientists in the following broad areas: (1) Science must establish more firmly the public-health basis for estimating the benefits of pollution control. (2) Scientists should take a more active role in formulating public policy and should establish an advocacy position in favor of more enlightened regulation based on as much scientific information as can be obtained through research.

THE PROBLEM

There was little disagreement among the panel members that some form of pollution-control regulation is necessary. As one participant put it, such regulation "legalizes industry's moral sense:" it enforces the sort of

Merril Eisenbud is Professor and Director, Laboratory for Environmental Studies of the Institute of Environmental Medicine at New York University Medical Center.

0077-8923/79/0334-0121$01.75/2 © 1979, NYAS

public-spirited behavior a firm could otherwise engage in only at the expense of its competitive position.

Nevertheless, there was a clear consensus that the flurry of Federal pollution-control regulation during the past decade has resulted in a dramatic overcompensation for past laxity. The pendulum has swung too far.

Indeed, while the costs of this recent Federal regulation are becoming increasingly clear, its benefits remain far less apparent. The panel considered the case of sulfur dioxide abatement in New York City. In the years between 1964 and 1969, the city reduced ambient concentrations from 0.2 parts per million (ppm) to 0.06 ppm by mandating the burning of lower-sulfur (and thus more expensive) fuel oil for building heating. This dramatic reduction was bought at a cost in the neighborhood of $80 million annually (in 1970 dollars). In 1970, the Federal government imposed standards on the sulfur content of fuel oil that had the effect of lowering the permissible ambient concentration of sulfur dioxide from 0.06 ppm to 0.03 ppm; and, although this reduction was tiny by comparison with the earlier reduction, it exacted an additional cost of about $200 million per year (in the same 1970 dollars). This $200 million per year undoubtedly represents one of the largest investments ever made for public health, yet the 0.03 ppm reduction in sulfur dioxide results in no measurable improvement in public health.

The panel laid the cause of such regulatory excess to a pathology of perspective: the tendency to consider pollution control apart from its full economic, environmental, and historical context.

Pollution-control regulation is economic policy. The cost of pollution control affects the price of goods manufactured by American industry as well as the nature and extent of new investment by industry. This has a very direct effect on America's competitive position in international trade. The United States is becoming less and less an economic island unto itself, and must make economic policy, including pollution control policy, with increased attention toward its effect on the country's export capability.

Pollution-control regulation is also environmental policy. But it is only one aspect of environmental policy. The panel emphasized the need for a much broader definition of "the environment" that would encompass all facets of the quality of life, and, concomitantly with such a broader definition, a more balanced environmental policy. Do not such other aspects of public health and human welfare as housing quality, better recreation, rat control, public-health education, and so on, merit a more equitable share of the national investment in "the environment?"

One must also consider pollution control in historical perspective. As example of sulfur dioxide abatement in New York City shows, it is a misperception to think that effective pollution control was entirely lacking before Federal intervention, and it is also a misperception to think that no Federal pollution control activities existed before the legislation of the last decade. Indeed, one of the effects of the recent reorganization has been to destroy much of the professional expertise that once existed in the Federal government: the highly professional and quasi-military Public Health Service, which once had the Federal pollution control mission, was moved to the Department of Health, Education and Welfare, and its powers subsequently transferred first to the Department of the Interior and then to the Environmental Protection Agency. The original professional corps, and its professionalism, did not survive the moves.

In this regard, the panel frequently contrasted the American pollution-control system with the British system. The latter, which evolved from the Alkali Act of 1861, employs a network of highly professional "Alkali Inspectors" with broad powers to enforce pollution abatement according to a "maximum practicable control" principle.

Perhaps the largest historical misperception is the notion that environmental health effects have been on the increase during this century. The incidence of the major forms of cancer, for example (excluding lung cancer, which is certainly caused far more by cigarette smoking than by environmental pollution), have actually been on the *decline* during the last 40 years—a trend that started well before recent Federal legislation.

THE RECOMMENDATIONS

The central emphasis of the environment panel's recommendations might best be put this way: yes, we need more research in the area of pollution control, but the more important need is to get the ideas and views of scientists into the political and regulatory system in an effective way.

The group suggested that academic scientists have heretofore been remiss in their insufficient representation of science before Congress and the Executive agencies. Indeed, the scientific community should form what one participant described as "a lobby based on logic."

The panel discussed a number of specific reform proposals, including the following: (1) Decouple the research function of federal pollution-control agencies from their regulatory function. (2) Establish a regulatory budget for the Federal government that would require explicit calculation of the indirect costs imposed on society by regulation and would set a ceiling on such costs.

In the area of research, scientists also have an urgent task: to establish a firm scientific basis for calculating the health benefits of pollution control. The panel suggested that many rules are now promulgated and many standards set with inadequate information as to what the benefits might be. And whenever scientists do not have a clear understanding of the public health effects of environmental pollution, it provides a weapon to those who seek to regulate with no attention to cost.

The panel also set out a research agenda for social scientists. The first task is the development and popularization of a sound risk-assessment and cost/benefit/risk-tradeoff methodology. The second task reflects the panel's general astonishment at the apparent complacency of the voting public toward such costly regulation. If this is a society regulating itself to death for no good reason, the group averred, it is a job for social scientists to find out why.

Communications

J. E. GOLDMAN

HAVING SPENT a part of my professional life in the field of transportation, I cannot help drawing some analogies between the social histories of transportation and communication as the communications industry approaches the same degree of maturity. These are more than analogies: communications through the history of mankind until the invention of the telegraph did, in fact, go hand in hand with transportation. Except in the case of smoke signals, communications began with transportation. Early attempts were quite dramatic, such as the 25-mile run in Ancient Greece to bring news of victory. Horseback, and most notably for a brief period in history the Pony Express, continued the tradition of information dissemination by physical effort. Trains and telegraph arrived approximately simultaneously. Trucking, automotive, and finally air transportation continued the evolution. Even in the face of all of modern communications technology, the belly of a Boeing 747 has the highest bandwidth of all communication carriers. If loaded with microfilm, its capacity is roughly 10^{12} to 10^{13} bits per second. If optical disks are substituted, this is increased by another two orders of magnitude.

Returning to the analogies between transportation and communications, let us look for the moment at how governments came into the picture. At first, anyone could have a horse and travel as he pleased and where he pleased, and even into the twentieth century there was little inhibition or control placed on the use of horses—not even on the pollution resulting therefrom. As the density of horses or conveyances powered by horses grew, local jurisdictions began to create roads and took responsibility for the care and maintenance of those roads. As transportation evolved into more complex modes, regulation became endemic. Private tolls were an early form of rate setting. Standards and protocols became essential, such as passing and stop-and-go conventions. As railroads became ubiquitous, government involvement increased by way of land grants, right of way, eminent domain, rate making, route licensing and safety standards. The advent of automobiles, trucks and air transport

J.E. Goldman is Senior Vice-President and Chief Scientist of the Xerox Corporation and a member of the Board of Directors, Xerox Corporation.

0077-8923/79/0334-0125$01.75/2 © 1979, NYAS

brought even more intensive government regulation extending to standards and operating procedures as well. Moreover, air transport brought the need for international inter-governmental agreements and jurisdiction. It might be noted in passing that the initial involvement of the United States government in all aspects of air travel for the purposes of assuring safety and stimulating research into improved transport, led to the creation of the National Advisory Committee for Aeronautics after World War I, which was singularly responsible for the great technical advances gained by the United States in aeronautical technology.

The trend of government involvement as the automobile became a dominant mode of transportation is very similar: at first there was only minor involvement through state licensing of automobiles, but it intensified with the years as vehicles and roads proliferated, leading to imposition of standards through inspection procedures and culminating in the extensive regulation of safety, emissions and fuel economy that we now see.

Patterns of regulation in the field of communications have followed a very similar trend. Regulation of the rights of way and of rate making for both telegraph and telephone communication came first. Then as radio communication (and broadcasting) burst upon the scene, it became necessary for the government to control the channels so that everyone utilizing the electromagnetic spectrum did not get in each other's way, a measure that ultimately led to complete regulation of channels and rates and established the principle of pricing to value rather than cost in order to protect the huge investment originally established to enable the most elementary of communications.

We are now entering a period in which a mood of deregulation has gripped the country. Mail service has been returned to private enterprise, the airline industry has been basically deregulated, and other forms of transportation are in the process of being deregulated. This trend is now spreading to communications. Three separate bills in Congress are focusing on it. The issue is not one of regulation versus deregulation, but the definition of that portion of the communication business that is to be deregulated, and of additional required regulation.

Technology and regulation combined to create the major impetus to reduce regulation. Regulation of those established routes where costs were low in comparison to revenue provided the business opportunity, while technology provided lower-cost implementation, to compete with the existing plant. These two forces, however, complicate the very issues they are intended to resolve.

The newer technologies and the expanding technologies which have mushroomed in the last two decades in the field of communications have given a sense of new dimension to what is possible, what is practical, what is practicable, economical, and equitable—considerations that I think have to be very seriously entertained when one tries to determine what makes sense in the communications field.

Now, what are some of these new and expanded technologies, and what are the effects they have? Fiber optics, satellites and cable systems have come along rather well, assuring virtually infinite bandwidth. If we add packet switching, contention access (such as ETHERNET), VLSI, microprocessors, digital communication and distributed switching, we can forsee optimization of channel paths right at the terminal with audio, video, data and raster inputs of hard copy all piggy-backing on each other.

In light of this, what are the key issues of science and technology policy that one must face? The first is one already mentioned. It is not a technological one but a quasi-political one, namely, regulation versus deregulation. The political dimension is manifest in the need for, and the traditional role of government in, preventing chaos—a possible eventuality if the time is not ripe yet for deregulation. But the technological component is more than just an overtone in this issue. This is an industry that has been led by technological forces. Traditionally, competition has spurred technology. But in some ways technological advances in communications have not suffered by the guaranteed monopoly that has been the more widespread experience of communications in the United States. So what are the significant issues that will ultimately determine an overall policy?

Value versus Cost

Telephone services have been historically priced on value, not costs. A subscriber living 300 miles from the nearest switching office pays the same rate as one who lives next door. A long-distance call using the heavy traffic routes between New York and Chicago is charged the same as a call between Holmdel, New Jersey and Naperville, Illinois. A subscriber who receives his service from a modern switching center pays the same rates for equivalent services as one who receives it from the obsolete step-by-step systems.

Deregulation means competition, and for competition to operate effectively, prices must be based on cost. Dramatically increasing the rates in rural areas or between relatively remote geographical locations, or as a function of the value of the switching office is not politically viable. Telephone services are viewed as a right and a necessity, and it is unlikely that such discrepancies will be tolerated.

INTEGRITY OF THE NETWORK

The structure of regulation utilizing the rate base encourages longevity in installed equipment and does not directly minimize investment. As a result, the incentive is for high quality. Additionally, customer awareness of lower quality would definitely influence state regulatory-commission charity towards requested rate increases. A depreciation life of 20 years would be in jeopardy if components fell apart after three years. The substance of regulation enhances the integrity of the plant.

In a competitive environment, this is not necessarily the case. Many of the telephone instruments available today from retailers do not function effectively, exhibit severe problems in dialing, cannot withstand impact, and often cease to function after a short time.

A network can generally be said to be as strong as its weaknsest link. A mechanism must be found either to identify problems arising from inferior products or to assure that the network itself does not deteriorate because of poor performance of a segment. A regulatory process for evaluating and approving equipment may be necessary.

VALUE OF PLANT

The value of the existing plant is the major element of rate making. The depreciation interval and the mode of depreciation are extremely important parameters influencing investment. It is likely that competition would insist on different depreciation rates and have valid reasons for demonstrating the reasonableness of those rates. The dramatic decrease in the cost of electronic componentry is certainly one force that suggests a shorter lifetime. A major portion of the plant unfortunately does not consist of modern technology and thus could not effect a major change quickly. All competition should be able to compete on the same basis. Yet, existing communication carriers might not be able to do so.

IMPACT OF TECHNOLOGY

With the advent of fiber optics, an historic trend is overturned. Fiber optics represents a revolution, not an evolution. In the late 1890s, telephone lines were first constructed carrying one voice per twisted pair; by the mid-1920s, carrier telephony was first used permitting 12 voice channels on one wire pair; in the later 1940s, coaxial cable was developed carrying 600 voice channels per conduit; in the 1970s, microwave links carry 1,800 voice channels per conduit; and in the early 1970s, communication satellites with tens of thousands of channels were in use. In each case, technology provided a lower cost per mile. It was obtained by concentration of the number of voice circuits in the conduit. Fiber optics will overturn this evolution. It has, in essence, made wave guides obsolete. It is about to be competitive with coaxial cable, then TI carrier and by the late 1980s the economics of copper wire will be surpassed by those of fiber optics. The parallel decrease in the cost of microelectronics is a second ingredient. This thrust of technology, if nothing else, would cause a rise in competitive pressures. It clearly threatens the value of the plant and offers unparalleled bandwidth availability.

Switching centers were originally constructed to optimize expensive transmission routes. In the future, inexpensive transmission routes will be constructed to optimize switching. The entire extant plant will have to be reevaluated.

Response to Pressure Groups

The Antidigit Dialing League was one of the first organized citizens groups protesting against the communications industry. Rate adjustment always attracts interest but in a more scattered way. The recent microwave-pollution allegations, satellite re-entry problems, and local requirements for underground buried cables are clear examples of counteracting pressures. It is inevitable with fiber optics that the Environmental Protection Agency and the communications industry shall cross swords. Fiber optics requires right of way, right of way requires access, access requires destruction of the environment. Though potentially not as serious as the Alaskan pipeline conflict, the issue is clear. The communications industry will be subject to increasing pressures from groups whose interests conflict with theirs.

Communication as a National Asset

Communication networks have strategic political, military, and economic value. At present, the political impact of deregulation is overwhelming these values. In the future, however, as business and government become more dependent upon communication, its value as a national asset will increase and its need for protection will become vital.

The question of government ownership may again arise and portions of the communications industry may once again face nationalization.

Conclusions

The evolution of communications is spawned by revolutions in technology. This is influenced by costs and opportunities. The individuals responsible for, and acting as spokesmen of, the technological community now find themselves enmeshed in the politics of regulation and deregulation. More than ever before, a clear viewpoint with regard to technological alternatives is needed to guide political decision. Either politicians could become technologists or technologists could become politicians. The former notion seems somewhat enthusiastic, and the latter reactionary. Another alternative is that the two groups work effectively together. Technology and politics must become cooperatively involved with one another, and it would seem reasonable to suggest that the technologist take the initiative.

Acknowledgment

I wish to acknowledge the assistance of R. L. Brass, Director, Business Systems Analysis, Xerox Corporation, in the preparation of this paper.

Overview of Policy Issues: Panel Report

Panel Chair: LEE L. DAVENPORT
Panel Members: Lewis M. Branscomb, Robert L. Brass,
William D. Carey, Duncan Davies, John M. Dutton,
Martin C. J. Elton, Arthur Gerstenfeld, T. Keith Glennan,
Ralph Gomory, David A. Irwin, Donald D. King,
Mitchell Moss, R. R. Ronkin, A. George Schillinger,
John F. Tormey, and Charles Weiss
Rapporteur: Robert Stibolt

THE USE AND APPLICATION of technology cannot be separated from politics, economics, and societal values. Particularly in the field of communications, these factors constitute an intricate context that will substantially influence the type of technologies that are developed in the coming decades. How to assess more effectively considerations in this context such as fostering innovation, reducing costs, maintaining standards for reliability, increasing energy efficiency, ensuring environmental quality, revising regulations and protecting the public interest, is a complex process which is receiving considerable national scrutiny.

One issue concerning the compatibility of technology and societal values involves access to public media. For example, prerecorded video tapes and discs are likely competition for television broadcasting. This competition could raise the overall quality of programming, or at least allow for viable high-quality programming through tape and disc distribution. The growth of video distribution, more importantly, means that broader access to the television medium is achieved since one no longer needs a broadcast license to reach television audiences. This is an example of a situation in which competition could result in changes believed to be socially desirable—in a free society, access to information media should be as broad as possible.

The trend toward both increased competition and deregulation of communications is causing a broad reassessment to be made of many factors accepted in the past as barriers to change. A series of things arising from this trend can be identified:

1. Noncarriers selling valued-added services based on services pur-

---

Lee L. Davenport is Vice-President and Chief Scientist of the General Telephone and Electronics Corporation.

0077-8923/79/0334-0131$01.75/2 © 1979, NYAS

chased from common carriers (i.e., Telenet) should not be regulated. Furthermore, carriers should be allowed to compete in these markets with reasonable accounting separation of regulated and unregulated activities.

2. The rapid expansion in the "effective bandwidth" of modern communications systems (estimated by one panel participant to be about 17 percent per year) is creating possibilities for greater competition and pointing the way toward some deregulation.

3. Technological limitations motivating regulation in the past will decline in importance as technology advances and unit costs decrease (in real terms).

4. Competition associated with deregulation will undoubtedly lead to greater innovation and falling costs.

5. In a competitive environment, pricing for services will be based on the cost of providing those services rather than on the value of the services provided.

6. Cost-based pricing will mean that urban and business users of telecommunications services will no longer subsidize rural consumers. Thus the long-held goal of equity in pricing would not be abandoned.

7. Total deregulation is probably impossible because of constraints to preserve quality, reliability, energy efficiency, environmental compatibility, and the like.

8. One must also consider the broader issue of how a competitive communications system in the United States might interface with largely regulated systems in other countries. Some nations, such as Singapore, heavily subsidize their communications systems to attract business from all over the world. International standards are important in order to guarantee international compatibility of systems and thereby allow each nation to control what goes on within its borders without adversely affecting the rest of the world.

A fundamental issue that must also be raised is, "What is the true 'social value' of expanded communications capacity?" If the only result of additional bandwidth is more television programming of questionable quality, then, perhaps goals of increased "social value" will not be achieved. Whether in fact we need additional bandwidth and, more importantly, who should determine its use, requires careful evaluation. Several options may be appropriate. More attention can be paid to educating users in the creative use of communications technology. However, the ultimate determination of how communications capability will be used may reside most efficiently with consumers via the marketplace. Attempting to make this determination through any kind of political mechanism has already proven troublesome and can result in a misuse of valuable resources. For example, one special-interest group has been asking NASA for years to build a $40 million public service satellite to offer services that could be provided far more inexpensively by the private sector through other means.

By contrast with a general concern about policies for guiding the use of communications technology, there is essentially no concern about the underlying health of that technology. The potential and vitality of the science and technology in this field is outrunning the ability to make use of it. Certain technologies hold the promise for the reconciliation of once conflicting societal goals. For example, the distance insensitivity of satellites implies the possibility that rural and urban telecommunications services may be cost-based without abandoning equity pricing.

There are key technologies that are having and will continue to have major impact on the communications field. Each, in varying ways, can improve service, reduce cost, open new horizons and at the same time be employed in ways that will enhance societal goals: a particular instance is the widespread growth of digital communications both for voice and data. The tremendous impact of solid-state integrated circuits for both electronic memory and logic is another example. The discovery of practical optical communications techniques centering on cost-effective optical fibers is clearly the most exciting new step in broadband and high-capacity transmission methods.

Each technological advance creates both new opportunities and new problems. In a regulatory environment, where the employment of new technology is increasingly blurring the distinctions between regulated service and nonregulated activities, problems for regulators can be serious. Indeed, the introduction of some new applications for technologies have been delayed because of a lack of regulatory decisions.

A final observation should be made on the respective roles of "technology push" and "market pull" in innovation (the former sometimes government-induced, the latter customer-induced). There is widespread consensus that technology push cannot spur innovation in the absence of market pull. If anything, the majority of useful innovations are predominantly attributable to market pull. It follows, then, that no amount of Federally-sponsored research and development can be expected to keep innovation vigorous in the absence of a healthy user environment. Economics drives technology, particularly where the economic and regulatory climate is favorable.

The Economy: Introductory Remarks

ELMER B. STAATS

INTRODUCTION

A DISCUSSION OF science and technology policy and its relationship to the economy of the United States comes at a most opportune time because of the current concerns of policy-makers with the performance of the American economy and its ability to generate economic growth, employment, and stable prices. The current performance of the economy points to several areas of concern. Productivity gains have averaged only 1.6 percent during the last decade—a discouragingly low figure compared with the 3.2 percent average for the first two decades in the postwar period and the 5 and 6 percent figures of our major trading partners. The slowing of productivity growth in the United States economy for the past ten years has resulted in slowing of economic growth. If productivity during the last ten years had increased at the same 3.2 percent annual rate of growth of the two previous decades, then output per hour would have been 11 percent higher in 1977. The difference would have meant more than a $100 billion increase of our nation's output in terms of real gross national product.

In attempting to explain the slowdown of productivity advance in the past decade, economists tend to concentrate on four measurable factors: slowdown in the growth of capital stocks per worker, increasing proportions of inexperienced employees, changes in the industrial composition of employment, and declines in research and development (R&D).

The greatest hope for increasing the rate of productivity growth lies in advances in technological innovations resulting chiefly from organized research and development, and in increasing the growth in productive capital to keep pace with the growth of the labor force. There has been a failure to recognize that productivity growth is not only affected by the efficiency of labor, but also comes about by incorporating new and more advanced technologies, such as computer-aided design, into new business capital.

Elmer B. Staats is Comptroller General of the United States.

0077-8923/79/0334-0135$01.75/2 © 1979, NYAS

Growth of capital investment, which has lagged behind historical rates in the current economic recovery, and increased outlays for research and development, are critical both absolutely and in relation to the growth of the labor force if our economic fortunes are to improve in the years ahead.

A point of concern arising from any assessment of the economic outlook is the relative decline in research and development outlays over the past decade. These declines are most likely to have an adverse effect on the rate of productivity growth in the years ahead. For example, total R&D spending in 1977 is estimated by the National Science Foundation at 2.2 percent of the gross national product (GNP), compared with 3 percent in 1964. The United States spends more than half of its research dollars in defense efforts, whereas the bulk of research expenditures by other major industrial nations with better productivity records has been in nondefense areas.

In 1975, private industry employed 5 percent fewer scientists and engineers than it did in 1970. And the overall United States patent balance declined almost 47 percent from 1966 to 1975. Because of the importance of technological innovation to productivity and our overall economy, these indicators are of major concern to policy-makers.

Research developed by the National Science Foundation has concluded that the contribution of research and development to economic growth and productivity is positive, significant and high. According to the 1977 Commerce Department report, technological innovation was responsible for 45 percent of the nation's economic growth from 1929 to 1969.

When high- and low-technology industries are compared, high-technology firms have productivity rates twice as high, real growth rates three times as great, one-sixth of the annual price increases, and nine times the employment growth. The same kind of favorable ratio prevails in terms of international trade. The trade balance for R&D-intensive manufactured products has been generally rising through the period 1960 to 1976, and is now more than $28 billion. The trade balance for non-R&D -intensive products is down from a break-even level in 1960 to a $16 billion deficit.

While these trends show the importance of high-technology industries to the economy, their growth has been reduced drastically. As recently as 1968, 300 to 400 new high-technology industries were to be found. But in 1976, the number was zero.

There are few viable solutions to the current economic dilemma, but productivity growth is one of the economic solutions that would benefit all segments of society. Higher productivity enables workers to take home paychecks that would more than offset price rises, it is able to contain inflationary pressures, and it contributes to the economic growth of the nation.

Science and technology policy directed to increased productivity would have substantial economic benefits. Insofar as science and technology policy contributes to the productivity of our nation's resources, we have a potential solution to our economic woes and a way to lower prices and to improve our standard of living.

The Economy

ARTHUR M. BUECHE

IN EXAMINING the critical relationships between science and technology policy and the economy, we may profitably concentrate on those issues affecting the role of science and technology in providing the innovations that contribute to economic vitality. Industrial innovation, of course, includes new methods of management, finance, marketing, and distribution as well as technological innovation, and success often requires innovation in more than one of these areas. However, in view of the purpose of this Conference, it seems appropriate to focus our discussion on technological innovation. Also, to proceed with an identification of issues, I do not think it will be necessary to review the literature in order to reach agreement, from the beginning, that innovation can—and does—contribute to the achievement of national economic goals.

As you know, the work of Edward Denison and others has confirmed that innovation is a major source of economic growth; it can help control inflation; it creates jobs; it helps achieve a more satisfactory balance of trade; it is the single most important contributor to productivity improvement; and properly managed, technological innovation can contribute significantly to improving living standards—the quality of life—for humankind all around the world.

I recognize that there are those who seem to disagree about the value of more technology. There are those who seek their own version of comfort by retaining only the fruits of past invention that suit their particular needs of the moment, meanwhile suggesting that other people—including future generations—have no need for, or right to, additional options.

This is part of the environment in which we find ourselves as "policy advisers" today, but it would be a waste of time to argue about the wisdom of trying to make clocks run in reverse, even if such were possible. For myself, I agree with the philosopher who has said that "the main cause of the good old days is a poor memory."

Arthur M. Bueche is Senior Vice-President, Corporate Technology, General Electric Company and Member, General Electric Corporate Policy Board and former President of the Industrial Research Institute.

0077-8923/79/0334-0138$01.75/2 © 1979, NYAS

Let us turn, then, to the matter at hand: identifying policies that will encourage innovation in order to strengthen the economy. I find it helpful, when faced with such a complex subject, to step back and look at the overall picture in the simplest possible terms. Oversimplification is often dangerous if misused in identifying answers to problems, but it is very useful in identifying the problems in the first place.

Therefore, I would like to start by categorizing the elements of innovation, as it relates to economic policy, into a simple listing of six fundamental segments:

1. There is the science base—the collection of technical knowledge and know-how built up through the years and constantly in need of replenishment.

2. There is the inventor. Again, to avoid being sidetracked, let us not be concerned whether this is the old-fashioned individual inventor or the more modern inventive team. It does not really matter; the important point is that there must be good new ideas, and good new ideas are created out of the marriage of human ingenuity with the resources of the science base.

3. There is the entrepreneur, who according to Webster's dictionary is "one who organizes, manages, and assumes the risks of a business or enterprise." In our society, the entrepreneur generally achieves his goals through the organization or use of private industry. He and his company reflect the important fact that science and invention are only the initial steps in the innovative process. Innovation is an overall flow process that must encompass the full spectrum of activities by which new ideas proceed from the laboratory to the marketplace and find useful application.

4. There are the customers. If you prefer, we can list this item as the customers/public. The entrepreneur errs who views his customers only as buyers of his wares; they are citizens, taxpayers, realists, dreamers, workers, contributors, people.

5. There is the government. One might assume, albeit naively, that government's chief role in the process would be simply to establish a framework of law and tranquility in which inventors' rights are protected, in which entrepreneurs can coexist profitably and productively with other entrepreneurs, and in which the customers/public receive equitable treatment. Entrepreneurs and the public alike would, then, pay the government a modest fee for thus providing services that they cannot perform otherwise for themselves. That, as I said, is what "one might assume" about the role of government if he or she were taking a new, fresh, rational look at the process. Much of our discussion will undoubtedly dwell on what kind of mid-ground we might seek between the

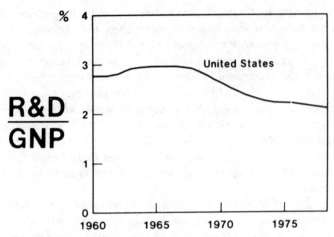

FIGURE 1. Research and development spending as a percentage of the gross national product in the United States.

idealized role for the government that I have just described and the entirely different activist/interventionist role that many others espouse.

6. There is the university. Our educational system, as characterized on my list as universities, is essential in many ways: as an all-important contributor to the science base; as the major source of inventors, entrepreneurs, and other skilled contributors to the innovative process; as a contributor to the attitudes of the customers/public; and as a source of policy advice to the government.

To focus on science and technology policy, then, means seeking ways to improve the interrelationships between these six elements, thus encouraging the creation and useful application of innovation for the benefit of all.

A look at the present environment—and current trends—in each of these six segments, provides ample reason to believe that some new policies are, indeed, urgently needed.

1. We are concerned, in the United States, about the strength of our science base. As shown in FIGURE 1, our research and development (R&D) spending, as a percentage of gross national product, has been going down since the mid-1960s. This is in contrast to what has been happening in some other major industrial countries (FIGURE 2).

Also, the ratio of R&D scientists and engineers in the United States population has fallen steadily since 1969.

Further, although it is harder to document, all of us involved in the

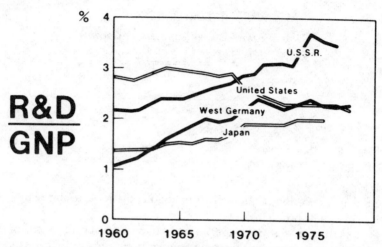

FIGURE 2. Research and development spending as a percentage of the gross national product in the United States, the USSR, West Germany, and Japan.

management of R&D have the distinct feeling that technical resources have been increasingly devoted to short-term evolutionary improvements at the expense of longer-range exploratory research.

2. A possible reason for our concern about the status of American inventiveness is found in the fact that the number of American patent applications—by American inventors—has fallen well below the 1970 level.

At the same time we note that an increasing percentage of American patents is being granted to inventors from other countries.

3. Of much greater concern not just to American entrepreneurs but to private industry generally, to those in government, and to economists, of course, has been the steady decline in the rate of American productivity growth. Note the figures showing the percent change per year (TABLE 1). It has dropped from an average of 2.4 percent per year in the 1950–67 period to only six-tenths of 1 percent in 1972–77.

Our share of the world market for manufactured products fell from 23 percent in 1960 to 18 percent in 1976. Meanwhile, the Japanese share, for example, was rising from 6 percent to 15 percent. And contrary to what might have been expected, the United States share of so-called technology-intensive manufacturers also dropped (TABLE 2).

4. As for for the customers/public, there is understandable caution and a certain amount of confusion over the current role of technology. Antitechnology and antigrowth voices are often loud and shrill, but

TABLE 1

UNITED STATES LABOR PRODUCTIVITY GROWTH
(REAL DOMESTIC PRODUCT PER EMPLOYED CIVILIAN)

Years	Ave. change per year (%)
1950–67	2.4
1967–72	1.1
1972–77	0.6

SOURCE: Joint Economic Committee.

there is evidence that the majority of the people look to innovation to help solve their problems.

As a matter of fact, overexpectation may be a greater problem. In the area of energy especially there is a feeling among many that if voiced, would say, "We needn't worry. Those scientists will come along with some miracle—solar energy, fusion, something—just in time and meanwhile energy conservation is a good idea for other people."

Also of long-term concern is the growing public desire for a completely aseptic and "risk-free" way of life. I think that we chance the loss of our national vitality and spirit if we divert too much of our resources toward the attainment of unreasonable, impossible goals. The tendency of the public to be less willing to take risks also manifests itself in a growing mistrust of entrepreneurs—and big business—where risk-taking is what makes the system work.

5. The average American's concerns about his government are simply that it is too big, too confusing, inconsistent, wasteful, expensive, and just plain ineffective. But such universal condemnation is not really very helpful. Trying to provide more constructive criticism—and specific policy suggestions—is, of course, a principal goal we are seeking to attain.

6. As for the sixth element in our model—universities (and our overall educational system)—the great current problem seems mainly a matter of financial support.

Inflation has hit hard at private institutions. Public schools have often been the first target of "taxpayers' revolts," simply because they are generally supported by the kind of local taxation over which the citizen has more immediate control. And the public institutions of higher learning suffer as their supporting governments devote increasing attention to the immediate problems of social distress rather than to the longer-term cures offered by education.

TABLE 2

SHARES OF EXPORTS OF WORLD MANUFACTURES

Country	Percentage	
	1960	1976
United States	23	18
Japan	6	15
Germany	18	20
France	9	8
United Kingdom	15	9

SOURCE: United States of Commerce.

These, then, are just a few of the current trends and environmental indicators for each of our six interrelated elements of the innovative process, the process we have identified as a key source of economic strength and growth. It is not a particularly bright picture; but then, if it were, this Conference would probably not have been convened.

The advantage of the kind of list I have constructed is that it permits us to seek out and identify the major problems without getting hopelessly distracted by a host of minor ones. Let us look at our list again.

We do need to rebuild our science base, but that is not a major problem compared with many others confronting us. Recent trends in government support of R&D reflect an encouraging new emphasis on basic exploratory work.

The inventive spirit may not be as unfettered as it was in the so-called Golden Age of Invention, but I see no real shortage of inventive capacity in our laboratories. And, if our vaunted "great American know-how" no longer reigns supreme in the world, we can rejoice in the fact that good ideas have never paid much attention to national boundaries in the first place, and modern communication can make new ideas from all parts of the globe more available throughout the world.

I also do not view current public attitudes—or consumers' changing wants and needs—as any insurmountable problem. Rather, I believe the desire for more environmentally compatible, more dependable, more useful, more value-packed products and services should be a challenge and an opportunity to the innovative process we are promoting, and not a threat.

As for the universities, their financial problems are indeed serious, and certainly there are many other issues that concern us. But the universities remain fundamentally strong and effective. The quest for excellence remains wonderfully evident. Academic freedom seems much less threatened today than it has been on occasion in the past. The new generation of students seems every bit as intelligent and motivated as we were in our day—even when we give those "good old days" the benefit of that not-always-accurate memory I mentioned earlier.

Our most critical and urgent problem areas are to be found in the interrelationships between the entrepreneur—or, if you prefer, the private sector—and government. But even here, in this highly complex area, a simplified list or model may help us focus our discussion more sharply.

Earlier, I described innovation as a flow process in which science and invention are only preliminary steps. To work, the process must include advanced development, engineering design, manufacturing technology, marketing, distribution, accounting—all the business steps leading to eventual application and delivery to the customer, who can put the product or service to good use. Each of these steps requires investment: of time, effort, skill, and money.

Risk-taking investment, then, is the pump that makes the whole innovation process work. This pump, you will note, does not necessarily operate at the front end—the technology end—nor exclusively at the other end, the marketplace. Rather, the entrepreneur often provides the driving force. He must be alert both to new technical capabilities as they come along and to the new needs and wants of the marketplace. His success comes with the interconnection of today's new technological capabilities with today's—and tomorrow's—new needs. (One of my business associates has suggested that it may be unproductive to spend too much time worrying about the relative strengths of so called "technology push" versus "market pull." As he says, "One man may start a fire because he wants to keep warm; another may start a fire simply because someone gave him a match.")

In any event, when businessmen are investing in new plant and equipment, they create a demand for advanced technologies for more efficient processes. Furthermore, we know that a very large share of the cost of bringing an innovation to the marketplace typically requires relatively large commitments to physical capital. Thus, general trends in capital formation, not simply trends in R&D investment, are the best indicators of the vigor of economic activity at the critical interface between science /technology and the economy. In other words, we can "'put some numbers" on how well that all-important "pump" is working.

Although 1978 was a relatively good year for business fixed investment, the rate of investment relative to gross national product (GNP) has failed to reach levels associated with past periods of economic recovery. Note in FIGURE 3 that business fixed investmment since 1964 ranged be-

REAL NONRESIDENTIAL FIXED INVESTMENT AS
PERCENTAGE OF REAL GROSS DOMESTIC PRODUCT
(1966–1976)

Country	Percentage of GDP
Japan	26.4
Germany	17.4
France	16.7
United Kingdom	14.9
United States	13.5

SOURCE: Organization for Economic Cooperation and Development.

tween 9 and 11 percent of real GNP. Investment increased in 1976 and 1977 but did not exceed the 1973 level until last year. Investment averaged 10.4 percent in the 1965–4 decade, but fell to 9.3 percent for 1975–7. The ratio rose to 10 percent for 1978, but this is still not adequate to make up for previous years. Further, a significant share of investment spending—$6 billion in 1977—was for installation of pollution-abatement equipment, which does not contribute directly to measured productivity.

By comparison, similar figures for other major industrial nations appear in TABLE 3, in which a very important comparison is shown. (Incidentally, the reason for the difference between the 13.5 for the United States and the 10.4 percent of GNP indicated in FIGURE 3 occurs because

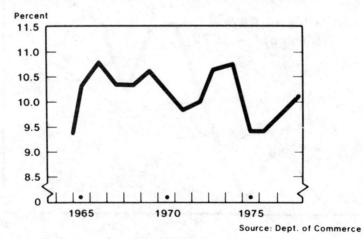

Source: Dept. of Commerce

FIGURE 3. Real nonresidential fixed investment as percentage of real gross national product.

the base for the OECD's figures on gross domestic product (GDP), shown here, exclude a variety of international transactions.)

Whereas the rate of United States investment has been—and if uncorrected, seems destined to continue to be—in the range of 10 percent of GNP, the Joint Economic Committee has estimated that achievement of national economic goals over the next several years will require an average of 12 percent. (To put this in perspective, this would have meant spending $90 billion more than we did over the past three years alone.) These goals include creating 11 million new jobs and reducing unemployment to 4 percent by 1983. The goals also include bringing inflation down to 3 percent and achieving a reasonable level of growth—approximately 4 percent—in real disposable income. Accomplishing this would mean regaining a level of productivity growth of 2 percent per year. You will recall that TABLE 1 shows that this growth rate had dropped to only six-tenths of 1 percent per year for the 1972–77 period.

All of us—technologists, businessmen, economists, and private citizens—are fully aware of the importance of cash flow, which is most conveniently defined as after-tax profits, less dividends, plus depreciation allowances. We all know that increased cash flow is one of the major requirements of increased investment, not only in R&D, but also in plant and equipment generally.

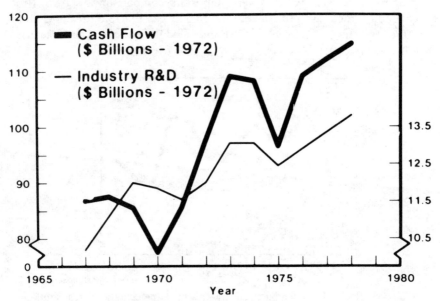

FIGURE 4. Relationship of research and development by industry and cash flow.

As can be seen in FIGURE 4, our R&D investment in the past has tended to move with cash flow. If this is generally true, we would conclude by regression analysis that the cash-flow change accounted for 84 percent of the variance in private industry's R&D spending since 1953. And the relationship with plant and equipment spending also was very close over the past 25 years. But now, in a new environment, the latter relationship seems to be changing. Notice the vertical distance between the cash flow and capital spending lines in FIGURE 5. Cash flow typically has exceeded capital spending by about $10 billion in recovery periods, but since the early 1970s the difference has doubled. Furthermore, current data on cash accumulation by American corporations suggest that there is cash on hand at this time.

Our key question today, then, is: *What policy changes are needed to spur longer-term investment—not just investment in R&D—but investment in all of the essential elements of the flow process of innovation?*

If businessmen have cash to invest, but are not investing it in new plant and equipment, there must be good reasons. I see no evidence that the current crop of executives and entrepreneurs are less courageous or less visionary than their predecessors. The reason, as a recent *Business Week* survey of industry leaders made clear, is "a profound lack of business confidence."

Uncertainty about the course of inflation and related Federal wage and price policies is the most severe problem, but government regulatory

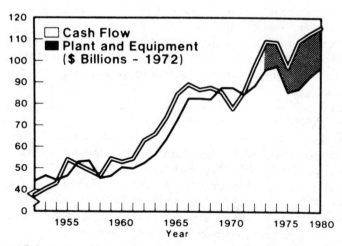

FIGURE 5. Relationship of capital spending for plant and equipment and cash flow.

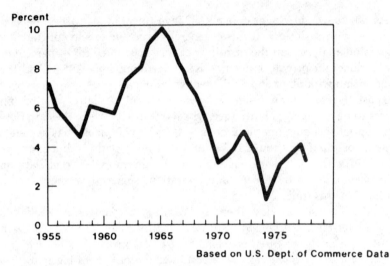

FIGURE 6. After-tax return on investment for nonfinancial United States Corporations.

uncertainty is almost as serious. Regulations can make or break a new venture, and the actions of government have become increasingly confusing, often impractical, and almost always unpredictable. Uncertainty about energy policy in particular is ubiquitous in its impact, not just on energy-related projects but on the very large number of investments for which energy sources and prices are important.

As another recent example of negative progress, there is the new Ethics in Government Act that creates more stringent conflict-of-interest rules, which are driving much of the best talent out of government service and reducing timely and meaningful access to industry experts on Federal-policy analysis.

But the risk/reward balance is the most important concern of those who would invest in the productive base of our economy. Unfortunately, the other half of the story is that while risks and capital costs have gone up, potential rewards have come down. Department of Commerce data show that corporate after-tax return on investment for nonfinancial corporations, adjusted for inflation, fell from 10 percent in 1965 to 4 percent in 1977 and to slightly below that in 1978 (FIGURE 6). The recent report of the Council of Economic Advisors analyzes four determinants of business fixed investment, and concludes that three of the four are well below the 1955–70 average.

Alan Greenspan explains the problem in terms of a risk premium having been added to the "hurdle rate of return," the rate that corporations set as a minimum in evaluating potential new projects. The risk premium is assigned to represent the potential for less favorable project costs resulting primarily from inflation, governmental energy policy, and regulatory policy. The fact that longer-term investments are hurting relatively more is cited by Greenspan as evidence that, due to a risk premium, the discounting of earnings in the distant future is very heavy. We see, in fact, a relatively greater shortfall in investments with longer-term payoffs, such as industrial plant and basic research.

What needs to be done? Business managers and economists who have concentrated on science and technology policy in the economy do not necessarily agree on the answers, but I believe there is reasonable agreement that policy actions are required to strengthen incentives and reduce investment costs and risks.

Let us now review checklists of some of the reasons for our nation's inadequate innovation rate. We might call them checklists of "investment disincentives" (TABLES 4 and 5).

First, however, it seems only fair and prudent to acknowledge that there are no easy answers. The policy-makers, or, in our case, the policy advisers, face a number of seemingly inescapable constraints and dilemmas. For example, in today's climate of worry about inflation, we obviously must be wary about proposing any new policy or guideline that will cost very much. Second, we must accept the political "fact of life" that there will be opposition to any compromise in environmental, health, or safety regulations for economic reasons. Third, existing knowledge and methodology for cost/risk/benefit analysis—and for assessing the relative efficiency of alternative incentives—have so far proved inadequate for making an airtight case for any policy changes we may recommend. Persuasions must still supplement logic, no matter how

TABLE 4

INVESTMENT DISINCENTIVES—STRAIN ON RESOURCES

1. Decreased rate of real business profitability (after inflation adjustments).

2. Income tax systems and high rates encourage consumption but discourage savings and productive investment:
   a. Rate increases, due to inflation, are faster than real income increases.
   b. Corporate profits are taxed twice if paid out in dividends.
   c. Tax on capital gains discourages risk-taking in productive investments.

3. Securities regulations constrain investment and investment liquidity in new business ventures; this aggravates capital shortage for new and small businesses.

TABLE 5

INVESTMENT DISINCENTIVES–CLIMATE OF UNCERTAINTY

1. Government regulatory actions:
    a. These are rapidly increasing and unpredictable.
    b. Regulators are poorly equipped to make objective and balanced assessment of costs (contribution to inflation, job loss, investment contraints, competitive disadvantage) versus benefits (health, safety, clean environment, maintenance of competition, etc.).
    c. Legal constraints and adversarial relationships restrict ability of government and industry to work out balanced solutions.
    d. Implementation is often inconsistent, confusing, and unnecessarily expensive, and projects are delayed by court actions.

2. Inflation:
    a. Inflation increases the discount rate on the future.
    b. It makes cost and profitability projections much less reliable.

3. Policy uncertainty:
    a. Energy (potential costs and government preferences for certain technologies).
    b. Foreign trade (diplomatic and national security controls, administrative delay).
    c. Monetary (interest rates).
    d. Antitrust enforcement.

pure we may judge the logic to be. With these provisions in mind, let me quickly summarize some of my own beliefs.

In a general way, perceived risks in the investment environment are affected by the level of business confidence regarding government policymakers' understanding of the true effects of their decisions (or "indecisions") on the economy and business management. There must be a reasonable expectation that policies crucial to long-term investments will remain relatively stable, or, at the very least, that changes will be practical and internally consistent. Demonstrated leadership in this direction by the Administration and by the Congress can have extremely beneficial effects, not only on the private sector, but also throughout the working levels of government.

Reduction of the debilitating "adversarial relationship" between the regulator and the regulated is similarly important. In addition to the impact of leadership from the top, the introduction of more opportunity for meaningful government/business interaction and personnel exchange can be very helpful, so-called conflict-of-interest laws permitting.

The inputs of the Industry Advisory Committee to the Federal Domestic Policy Review (DPR) on Industrial Innovation represent the most up-to-date indicators of industry views. The Advisory Committee's highest-priority recommendations are in the areas of regulatory reform and tax incentives. In the regulatory area, the DPR Advisory Committee has underscored these important issues: (1) how to assess costs, risks, and benefits in such a way that the public and policy-makers can better understand the alternatives and trade-offs; and (2) how to most efficiently accomplish the proposed goals of regulation, that is, how to take advantage of industry's innovative capacities to satisfy the environmental, health and safety needs of society. It should be noted that the Administration already has underway a number of important innovations in regulatory implementation procedures, such as the Regulatory Council, the Regulatory Analysis Review Group, and the improvements contained in the Executive Order on Regulation. The recommendations of the Domestic Policy Review are expected to provide impetus for still further improvements.

However, the policy-makers' dilemmas I mentioned earlier are still very real. Virtually any regulatory reform brings down the wrath of powerful special-interest groups, and the analytic tools for assessing relative efficiency, and for demonstrating costs, risks, and benefits of regulatory alternatives, are pitifully inadequate. But I doubt whether anyone believes that regulatory and other policy uncertainties will ever go away completely. In fact, I am very concerned that they cannot be sufficiently reduced on the time scale needed if we are to regain the kind of investment in technological innovation that will help us to achieve our national economic goals in the early 1980s.

While many policy-makers are convinced of the need for tax reductions to strengthen investment incentives for plant and equipment, they are faced with the dilemma that any reduction in revenues could, by increasing the federal deficit, aggravate inflation. And inflation, in turn, would offset any benefit that might have been gained by the tax incentive. Recent history shows all too clearly how this can happen. In any case, the Administration apparently believes the possible short-term negative consequences of increased inflation outweigh the long-term benefits of increased tax incentives.

However, it seems possible that the issue deserves a more creative approach. If the forecasters are correct, we are looking into the face of a small slowdown in 1980, and a further reduction in the already inadequate level of investment. I would pose the question whether this reduction might not be headed off by a tax incentive which, if properly designed, would produce net benefits from the very beginning. Why not think of phasing in a faster capital-recovery package, beginning in 1980? I am referring to such things as a five year write-off on equipment, and ten year write-off on buildings. Wouldn't the promise of increased cash flow and of higher rates, of return stimulate some spending right away? And if so, would the increased economic growth quickly bump revenues sufficiently to overcome the costs of the incentive? I hope I can get your candid reactions to this proposal.

The Domestic Policy Review Industrial Advisory Committee also has made a strong plea for improving the economic assumptions now used by the government to calculate revenue impacts of alternative tax policies. We know that the use of so-called "dynamic macroeconomic models" in estimating the impacts of the recent change in the capital gains tax were very influential in the decision that was reached. But we also know that there was great controversy over the assumptions used in the models, and there is still a great deal of debate among the experts about the appropriateness of the trade-off that was made. Some experts argue, in fact, that these models, which try to estimate secondary and downstream effects of tax changes, should not be used at all until better understanding of the economic interconnections is gained and more agreement is achieved about appropriate assumptions. Here is a case where some innovative technology itself is needed as an eventual aid to spurring the future of technological innovation.

Here is another point to consider. Numerous suggestions exist for the establishment of new institutions such as cooperative technology centers, centers of excellence, and so on, all for the purpose of increasing collaborative efforts between universities, industry, and government. This collaboration is to be in areas of science and technology of common interest. I believe there is substantial merit in this concept as long as the private sector plays an important role in program planning, resource allocation, and management. For example, such a mechanism might improve the advantage that the civilian economy can obtain from the vast amount of effort going into science and technology programs for defense and could have early and significant benefit.

We have now reviewed what I am sure is an oversimplified, but, I hope, still useful, list of the six basic elements of the process of technical innovation. We have looked briefly at the present environment in each area, and at some recent trends. We have suggested that the single most productive area on which to focus new and improved policies is that of providing better incentives for risk-taking investment. We have made a checklist of key points in the relationships between government and industry that must be considered by policy-makers and their advisers. We have acknowledged the dilemmas inherent to the work of today's policy-makers. And I have sought to get our discussion started by presenting some of my own personal views about certain steps that I believe need to be taken.

Acknowledgments

I wish to acknowledge contributions to this paper made by L. W. Steele, B. J. McKelvain, A. Pearlman, P. S. Welch, and R. N. Landon.

Comments

ELMER B. STAATS

To REINFORCE the concerns raised by Dr. Bueche in his presentation about the current state of the United States economy, let me emphasize the need to meet the national policy goal of full employment. To meet that goal, the United States must attain technological innovation on a scale that has not been achieved for over a decade. Since 1955, the number of people reaching working age has been increasing at an accelerated rate, and at present nearly twice as many new jobs must be created each year as we needed some 20 years ago. Without the underlying scientific understanding and primary technological developments, countless employment opportunities would not have been available in the economy. A single, basic technological change such as a transistor or the integrated circuit provides thousands of opportunities for application to computers and consumer electronics.

I would like to point up some of the actions that the Federal government can take to address these current economic concerns. These approaches could be a matter of priority for the Federal government.

1. To develop periodic assessment of needs to determine the nature and extent of public- and private-sector productivity problems.

2. To act as a facilitator in bringing together various groups on neutral ground to discuss widespread industry productivity problems.

3. To operate a productivity clearinghouse providing national and international data and knowledge on various aspects of productivity to all sectors of the economy. Particularly, we need to provide private industry with more knowledge as to developments in foreign countries that may have applicability to the United States, or that may contribute to our international competitiveness.

4. To promote a better understanding of all the factors affecting productivity, including human resources, capital, technology, research and development, transformation of knowledge into practical terms, and the importance of productivity to our national economy.

5. To provide for a periodic joint assessment by the Joint Economic Committee of the Congress, the Council of Economic Advisors to the President, and the Federal Reserve Board, of the productivity impact of

---

Elmer B. Staats is Comptroller General of the United States.

0077-8923/79/0334-0154$01.75/2 © 1979, NYAS

fiscal, monetary, tax, and regulatory policies on the private sector of the economy.

6. To take the lead in developing improved and acceptable measures of productivity. Our current productivity statistics are weak and do not adequately reflect the role that capital investment, improved technological processes, and innovation can play in improving productivity. The Bureau of Labor Statistics and the National Academy of Sciences have done good work in this area, but much more needs to be done.

7. To adopt policies that will stimulate additional investments in R&D by the private sector through tax and other incentives, and to encourage industry to recognize the importance over the long term of R&D, rather than focusing on investments that will yield high short-term returns. The 1978 tax bill will help, but the question is whether it goes far enough. Extending the investment tax credit specifically to research and development outlays might provide further assistance. It is hoped that the domestic policy review of industrial innovation will result in a new cooperative approach to industrial innovation.

8. To provide new and better ways for measuring the costs and benefits of both existing and new regulations that can impact productivity. The Regulatory Analysis Review Group established by the President to review selected new regulations is a step forward, but the entire regulatory process needs to be subjected to the rigorous discipline of costs-and-benefits analysis, particularly with an eye to those regulations that have been designed to deal with health, safety, and the environment.

9. To continue Federal labor/management cooperative programs for upgrading the skills of the labor force, with added emphasis on service trades, which now make up some 60 percent of the labor force, and which are expected to grow to 75 percent by the end of the century.

10. To accelerate the efforts of the Federal government to measure and improve productivity within the Federal government itself, and to take a strong leadership role in assisting state and local governments to improve productivity and reduce their costs. A recent study estimates that 20 to 30 percent of state and local employment growth between 1967 and 1976 resulted from low productivity. Underscoring the importance of this point is the fact that state and local governments now employ 80 percent of all government employees in the nation.

I would like to conclude by saying that perhaps one of the best examples that comes to mind with respect to government/industry/private cooperation is in the field of agriculture, where we find that the Department of Agriculture, working with the American farmer over the many years, has created one of the most productive agriculture industries in the world by developing joint mechanisms for rural development, rural electrification, worldwide marketing and commodity programs, plus a host of others, including capital formation, and unquestionably the most effective R&D base and technology-distribution channels. American agriculture feeds not only the United States, but also a major portion of the free world. In fact, we see this model copied over and over in other nations, which in many instances have expanded the application to their manufacturing base as well. No doubt this has contributed to their more advanced productivity growth rate.

It is encouraging that there is a new interest in the subject of technology innovation. Although technology innovation does not in and of itself provide the solution, it is a basic ingredient for improving the nation's standard of living and promoting economic growth. Coupled with appropriate monetary and fiscal policies of the Federal government, technology innovation can be vital to the future of the country.

Overview of Policy Issues:
Panel Report

Panel Chair: ELMER B. STAATS
Panel Members: Alfred E. Brown, David N. Campbell,
Donald W. Collier, Duncan Davies, Bernard Delapalme,
Dennis Dugan, Merril Eisenbud, Arthur Gerstenfeld,
Donald D. King, Charles F. Larson, John Logsdon,
Mary Mogee, Rolf P. Piekarz, Geoffrey Place,
R. R. Ronkin, and Max L. Williams
Rapporteur: Ted Dintersmith

Many Problems Exist in the Economy

Even a cursory review of current economic indicators suggests that the economy of the United States is troubled. High inflation rates, unfavorable balance of trade deficits, and sagging productivity are telling statistics. Science and technology policy, by encouraging the process of innovation, has the potential to overcome some of these economic difficulties.

The productivity growth of the United States labor force has declined from an average percentage change per year of 2.4 percent in the 1950-67 period to 1.1 percent in the 1967–72 period and, most recently, just 0.6 percent in 1972–77. In contrast, many of the United States' main trading partners have been enjoying a 5–6 percent rate of productivity growth in the last decade.

The United States balance of trade deficits, due in large measure to petroleum imports, have been very large in the last five years. Most disturbing, however, is the fact that the United States now imports more technology products than it exports.

Innovation Is A Key to Economic Growth

Correlative evidence for the relationship between innovation and economic health is available and convincing. One signal of innovation—United States patent applications by American inventors—has declined markedly since 1970. In the 1967–1975 period, the United States patent balance has declined by 47 percent.

The conclusion that declines in American inventiveness and innova-

---

Elmer B. Staats is Comptroller General of the United States.

0077-8923/79/0334-0157$01.75/2 © 1979, NYAS

tion cause economic difficulty is suggested by the economic literature. One particularly relevant study by Edward E. Denison shows that from 1927–1969 technological innovation alone was responsible for 45 percent of the growth in the economy of the United States. In sum, innovation and economic health are highly interdependent; if science and technology policy is to contribute to the economic wellbeing of the United States by reducing trade deficits and inflation and increasing productivity, it must facilitate the entire process of innovation.

The Innovation Process is Broader than Research and Development

In analyzing how science and technology can generate innovation, the temptation is to limit discussion to R&D expenditures. The fact that R&D funds, as a percentage of GNP, have declined in the United States at the same time productivity has fallen suggests a strong relationship between R&D and economic health. Moreover, a closer look at R&D indicates that shifts hve been occurring in the composition of R&D expenditures.

A large portion of the R&D done in productivity areas is done by industry; and in the past twenty years industry's expenditures on R&D have been roughly constant. In fact, recent statistics show that companies spend $18 billion on R&D in 1977, a 16.4 percent increase over the money spent in 1976. Thus, the declines in total United States R&D have come only in the government-funded areas of space and defense R&D.

While industrial R&D outlays appear to be at healthy levels, the number of patent applications has fallen. The general consensus is that, in spite of R&D, innovation has faltered, suggesting that an analysis of innovation must go beyond R&D issues. If science and technology policy is to prod innovation, it must address the entire innovation process. While adequate funding for R&D is critical, other aspects of Federal policy—such as an uncertain economic environment, conflicts in and costs of regulations, availability of venture capital, tax policy (particularly with respect to industrial investment), uncertainty in wage and price policy, and tentativeness in energy policy—are the factors impeding innovation and, indirectly, economic growth.

The necessary steps in the innovation process cover a spectrum of activities. In addition to vigorous and creative research, the basic components include development, engineering design, manufacturing and production technology, marketing, and distribution. A breakdown in any of these areas will retard the rate of innovation.

Obviously, each phase of innovation requires a commitment of expertise, time, perseverance, and funding. A corporation deciding whether to devote these resources to innovative ends must balance the potential benefits against the immediate costs. Moreover, the success of a project hinges upon favorable outcomes in a number of areas; hence, an investment of corporate resources in innovative projects must be made in the face of uncertainty.

Because investment in innovation yields uncertain results, an entrepreneur committing funds to innovative areas takes risks. Investment in innovation—the essential element in the innovation process—must produce relatively high anticipated yields for such a project to be pursued.

Innovation and Capital Formation

One aspect of investment that is central to the innovation process is capital formation. Relatively sizable investments in physical capital are generally needed to develop and commercialize an innovation. Consequently, it is expected that general trends in capital investment will be an important reflection of the strength of the innovation flow from research to marketplace.

The statistics for the United States economy with respect to business investment suggest that this area is one of concern. Since 1964, business fixed investment has varied between 9 and 11 percent of real GNP. Importantly, the average investment, as a percentge of GNP, was 10.4 percent in 1965–74, but fell to 9.3 percent for 1975–77. It is noteworthy that an increasing share of business investment—$6 billion in 1977, for example—goes to the installation of pollution-abatement equipment, which does not contribute directly to measured productivity.

In contrasting United States investment rates to those of the United States' major trade competitors, interesting results appear. In the decade from 1966 to 1976, Japan's investment rate as a percentage of gross domestic product, was almost twice that of the United States. Germany's rate was 1.3 times that of the United States, France invested at a rate 1.24 times that of the United States, and the United Kingdom invested at a slightly higher rate.

Thus, in summary, the innovation process consists of a number of steps, all of which require financial commitments. In addition to requiring venture capital, investments in innovation involve risk. Corporate uncertainty about future government policy increases the uncertainty of innovative projects. Consequently, a Federal policy that seeks to promote innovation should adhere to two general guidelines: (1) increase the availability of venture capital, and (2) create a climate of regulatory stability conducive to "business confidence."

The availability of venture capital is particularly critical for small corporations. A large corporation, in some instances, will be able to generate internally the capital needed for investment in an innovative project. In contrast, a small business concern may have considerable difficulty in obtaining the funds needed to carry an innovative idea past the research stage and to the marketplace.

Innovation Issues that Need to Be Addressed

In general, the relative abilities of small and large firms to innovate is not well understood. A greater understanding of the relationship between the size of a firm and its ability to innovote should help to guide any decisions made, for instance, with respect to antitrust policy. While it is clear that long-term economic growth is contingent upon innovation, the debate about how the concentration of firms in an industry affects this growth is unsettled and ongoing.

Questions about how firms in an industry interact with respect to innovation may be important in identifying policies that stimulate economic growth through increased innovation. Arrangements under which several firms could conduct cooperative research have several advantages, including the reduction of overlap in R&D programs and the achievement of greater economies of scale in R&D. These advantages must be weighed against disadvantages, such as reduced competition. Since other countries have different approaches to interfirm competition and cooperation, it is possible that international data will convey useful information to domestic policy-makers.

In addition to examining interfirm issues, science and technology policy-makers should encourage research concerning the effect of a firm's organization on its ability to innovate. Social science research can yield insight into the relationship between organizational structure and institutional creativity. Questions to be addressed include: How should a firm organize its R&D effort?; how should a firm's R&D activity be integrated into the rest of the firm's activities?; and how should the effectiveness of R&D programs be evaluated?; how can a firm coordinate its various R&D activities?

The question of organization and innovation is not limited to the private sector. The effect of the government's organization of its R&D agencies upon innovation is, as yet, an unresolved issue. Topics that warrant further scrutiny include the degree of decentralization in Federal R&D decision-making, criteria to evalutate the effectiveness of Federal

R&D managers, the effect of Federal organization on risk-taking in R&D investments, and the relative availability of funding to particular sectors and to multidisciplinary areas.

A large organization can be more creative and innovative if its constituent elements, both management and labor, work cooperatively in a productive fashion. Teams involving both labor and management can best deal with the intricacies of R&D, engineering design, production technology, marketplace needs and distribution. By working jointly, labor and management can create an innovative environment. At the same time, cooperative effort will make each group sensitive to the hardships that particular innovations will place on the members of the firm.

The theme of cooperation underlies a proposal to establish cooperative technology centers. An environment in which people from government, industries , and universities could work cooperatively on our society's most challenging technological problems could make a profound contribution to our nation's ability to innovate. Creating multifaceted teams to attack science and technology problems would allow different perspectives and erspective and skills to be integrated would allow different perspectives and skills to be integrated in a single problem-solving unit.

In addition, people from different sectors of the nation, by working together would gain greater appreciation for each other's points of view. The current malaise of the "adversarial relationship" between business and government might subside if these groups shared the same information base; government/university/industry centers have the capacity to broaden existing information bases. Perhaps more importantly, the creation of these centers might signal that the nation as a whole recognizes that a substantial effort is required to further develop our science and technology strength, and that cooperation is the key to future success.

Independently of their role in these centers, universities make a substantial contribution to our nation's science and technology based and, indirectly, to American economic productivity. The manner in which university research is funded is a matter of concern. Recent measures of the Administration to provide increased funding levels are a crucial first step. Additional funding of research could yeild substantial long-term results.

While the absolute level of funds going to universities is a topic of great concern, the way this funding is disbursed is equally important. A situation in which academic research teams must sacrifice intellectual flexibility and creativity in order to support themselves with contract-based funding is not advisable. Dedication and commitment within the university system has not waned in recent years; consequently, grants that allow promising ideas (perhaps undiscovered at the time of the grant application) to be pursued are a sound national investment.

In conclusion, it appears that the way to approach the current "stagflation" in the United States is not only through monetary policy or fiscal policy, but monetary and fiscal policy in combination with science and technology policy. Federal initiatives to promote the entire innovation flow are imperative. The cornerstones of this policy should be a stable regulatory environment and a tax policy that ensure the orderly and appropriate availability of financial and venture capital to industry.

Regulatory Policy

MONTE C. THRODAHL

INDUSTRY IN AMERICA has developed, to a large extent, because of a progressive technical climate. Indeed, it is essential that this progressive climate—one in which science, technology, and innovation can flourish—be maintained for the proper nurturing of a resourceful and vital industrial sector.

More recently, however, United States industry often faces inconsistent, fuzzy, and sometimes unreasonable regulatory actions. Such regulations often result in unproductive diversion of our limited capital, and at the same time discourage research and development efforts.

The preambles of any number of regulatory laws leave no doubt that others, like ourselves, are concerned about protecting our environment, while still defending our economic system. But only when we see the legislation codified into regulation do we understand the full implications of restrictive regulatory provisions.

What can we do about regulations that begin with good intentions but become restrictive obstacles? What is required is a new focusing of regulatory policy, a more explicit definition of the means to reach our goals.

One indication of the health of our technical society is the level of innovation. We can all recite the various indices that point to a slackening of innovation in the United States. For example: (1) Currently about 35 percent of all United States patents are granted to non-American inventors; in 1961 the figure was 17 percent. (2) The rate of new-product introductions in pharmaceuticals is only one-fourth of what it was 20 years ago. Can this slackening in innovation be reflective of our attempt to follow regulatory blueprints that are defective? It is just this topic, the impact of governmental regulatory policy on science and technology, that is the subject of this paper.

Monte C. Throdahl is Senior Vice-President, Monsanto Company and a Member of the Board of Directors, Monsanto Company.

0077-8923/79/0334-0163$01.75/2 © 1979, NYAS

I am certainly not opposed to regulation per se. Regulation *can* serve as a catalyst. While achieving its intended social goals, regulation has the potential to improve profits, reward innovation and promote progress. But all too often, we the regulated find standards of compliance that are dictated by a confusing code of expediency with little regard for reasonableness. Such a policy can lead nowhere but to uncertainty for the regulated.

Many of the examples and policy recommendations that I will discuss were developed by a recent Domestic Policy Review Advisory Subcommittee of the Commerce Department on which I served. This subcommittee was looking into governmental impacts on innovation. On the basis of our findings, plus those of other industrial subcommittees, government agencies, and other interests, a set of focused recommendations is being prepared by Jordan M. Baruch, Assistant Secretary for Science and Technology, which will then be sent to the President.

Since the impact of regulation on technological development is so extensive, let me stake out the boundaries. I am going to exclude the impact of antitrust and price/entry regulations. Although these are important aspects of the government's total regulation, they are subjects unto themselves. I have also aimed my comments at the climate of regulation in the United States. While the examples are domestic, the principles of regulatory policy impact are universal. And before evaluating probable trends in regulation and discussing problem-solving approaches, let us look at where we stand.

Four problem areas of regulation are of concern. Much of the regulation today is detrimental to innovation because: (1) it is inconsistent; (2) it causes delay in adopting innovation; (3) it is inflexible in administration; and (4) it places more emphasis on the means of compliance than the results to be achieved.

Contributing to these growing regulatory requirements is a long-existing force wrapped in a new package—the force of uncontrolled and undisciplined single-issue advocacy. William J. McGill, President of Columbia University, recently noted that as the number of constituency groups increases, more dramatic ingenuity and more flamboyant rhetoric must be developed in order to attract headlines.

Uncontrolled single-issue advocacy is creating substantial fear and anxiety in society. This fear comes from the nightly cancer scares on the television news and movie and television dramas or documentaries that deliberately distort facts. Anxiety is promoted by the publicity releases of politicians and environmental groups, who are quite effective in securing the attention of mass audiences. But such actions can also be very dangerous.

The use of undisciplined advocacy comes from some surprising sources, the United States Senate, for example. Senator Edmund Muskie, speaking at the University of Michigan in mid-February 1979, vigorously expressed his perception of those who criticize environmental regulation as it exists today. He said "the environment is under attack from people who only believe numbers and figures."

Our "adversary society," as McGill describes it, is precious and must be preserved provided it does not become an end in itself. To my knowledge, undisciplined, uncontrolled advocacy has never been so extensive.

Let us now turn our attention to the specifics of the four areas of regulatory concern that I mentioned earlier.

INCONSISTENCY OF REGULATIONS

One of the most striking issues that surfaced in the various Domestic Policy Review Sub-Committee discussions on innovation was the inconsistency of government regulations. The chemical industry currently faces the handicap of attempting to comply with four separate cancer policies. These policies have been established by the Environmental Protection Agency (EPA), the Occupational Safety and Health Administration (OSHA), the Federal Drug Administration (FDA), and the Consumer Product Safety Commission (CPSC). To achieve some kind of order, the Inter-Agency Regulatory Liaison Group has issued a draft umbrella policy on cancer. Such action sounds laudatory. However, it appears that the end result will be a policy that essentially lets each agency continue to implement its own inconsistent program while adopting the most restrictive provisions of each of the other agencies.

Changing objectives, varying standards, and the uncertainty in the methods used to measure compliance to regulations all have an impact on science and technological development. Many firms are reluctant to commit sizeable resources to research projects having a high-risk of regulatory involvement. The result is direct impact on the quantity and quality of research and development.

My experience in this area is particularly vivid since I witnessed the dampening effect that inconsistent regulatory policy had on Monsanto's technical community when its innovative plastic beverage container was banned. The FDA had set forth a final regulation specifying conditions for safely using a raw material called acrylonitrile (AN) to produce soft-drink bottles. This chemical had been used in food-contact applications for more than 30 years. After that approval, and after the FDA had earlier reviewed extensive data submitted during the container's development, Monsanto and a major soft-drink firm committed millions of dollars and hundreds of jobs to the introduction of the product.

The bottle had a highly successful introduction. A number of government officials termed the container a truly innovative product from an environmental standpoint. Shortly thereafter, however, an interim technical report on toxicity tests involving rats fed acrylonitrile in massive doses was released. Based on these results and reports than AN could migrate into the bottle's contents, the FDA quickly banned the product without a full-scale review of the animal feeding results.

Monsanto researchers, using the most sophisticated testing equipment available today, cannot find any trace of acrylonitrile leaking into the beverage under realistic conditions. But the regulators say that if the bottles were filled with acetic acid and stored for six months at 120 degrees Fahrenheit, infinitesimal amounts of the chemical could leak into the solution. Despite the remoteness of the risk, the regulators chose to ignore the bottle's considerable real and potential benefits and banned the bottle. The regulatory action cost approximately 1,000 people their jobs and the company incurred a loss of approximately $100 million. Of greater importance to the country is the effect such a sizeable loss of jobs and dollars has on future ventures of the industrial community.

The episode serves as an example of what can happen even when an organization has gone to considerable lengths to reveal its entire development records, data, and experience to outside environmental experts, as well as committing more than seven years to research and development (R&D) and testing.

Changing objectives or uncertain standards of regulation, as well as uncertainty in the methods for measuring compliance, can slow technological progress. When regulation by the same or different regulatory agencies is contradictory, or when standards or methods of measuring compliance are not stabilized for an appropriate time period, firms are not willing or able to accept the risk of committing resources to potential innovations.

Delay in Adopting Innovation

My second point concerns the delays regulations cause in the adoption of innovative projects.

A Presidential task force on energy resources, evaluating the requirements for environmental impact statements, claims that a major uncertainty is *not* whether a project should be allowed to proceed, but rather how long the project will be delayed pending issuance of an environmental impact statement that can stand up in court.

A recent example has been the attempt of Standard Oil of Ohio (SOHIO) to modify an existing gas pipeline to an oil pipeline. This innovative project would move Alaskan oil from the West coast to the Midwest. After five years and $50 million in expenses and the submission of more than 700 permits and applications, SOHIO has announced its intention to drop the project. I understand that this decision is being reevaluated but delays have increased the cost of the project to the point where it may no longer seem attractive.

John R. Quarles, former deputy administrator of the Environmental Protection Agency, says that it is a "virtual certainty" that anyone involved in planning major industrial plants in the United States will see Congress change the environmental regulations affecting those plants before construction starts. During the past decade, Congress has revised major environmental legislation every three to four years, and it takes at least that long to plan a major facility. In a survey conducted by Mr. Quarles for industry, the Clean Air Act was seen as the most important of the new requirements. Other restraints on industrial planning are the Clean Water Act, the Resource Conservation and Recovery Act, the Coastal Zone Management Act, the Powerplant and Industrial Fuel Use Act, and the National Environmental Policy Act. This series of seemingly separate regulatory programs has totally changed the outlook for getting needed approvals for new plants from that of even a few short months ago.

Delay in approval of new products is also a major problem. New-drug approvals now have extended the timetable from invention to market by more than 10 years compared to the time required in 1962. That is more than half the patent life.

Inflexibility of Regulations

Regulations are frequently imposed on an inflexible basis, regardless of their merit or applicability to a specific situation. Perhaps this is because legislators are often lawyers and lawyers insist on reducing everything to yes or no, black and white. Toxicology, on the other hand, is an inexact science.

The inflexibility of the regulatory process is aggravated by the tendency of agencies to overregulate. The new Toxic Substances Control Act contains a provision concerning the requirement to file premanufacture notices on new chemical substances. These must be filed at least three months before manufacture of the chemical begins. Congress spent five years constructing this section of the statute and the final version does not require the EPA to promulgate rules concerning premanufacture notification. Thus the law could have become self-implementing. The EPA, however, has taken the position that such an approach would be a short-run expediency which would turn into long-run ineffectiveness. Consequently, the agency has proposed long and complicated rules and has accompanied them with amazingly long and detailed forms that must be completed when one makes a premanufacturing notice submission. If promulgated with their present requirements, completion of the forms will be tortuous and extremely time-consuming. This will make compliance overly costly and burdensome and will have a serious adverse impact on new-product development. This is a classic example of over-regulation by a Federal Agency.

The 1970 Clean Air Act established very stringent standards for automotive exhaust emissions. The law called for a 90 percent reduction in hydrocarbons and carbon monoxide by 1975 and a 90 percent reduction in nitrogen oxides by 1976.

The United States automobile industry had massive capital investments tied to their current automotive technologies. This, coupled with the increasingly high costs of regulatory compliance, made it virtually impossible to switch to newer, innovative technologies in time to meet the mandated end result. Instead, the auto manufacturers were forced into incremental improvements in current technologies.

Contrast this with the experience of the Honda Motor Company of Japan. Aided by a cooperative, government-industry financial program, and unfettered by any massive capital commitment to existing technologies, Honda engineers poured their R&D energies into an entirely new engine.

Because Japan had no emissions regulations, Honda had the time to refine and perfect that new engine in their domestic market. By the time of its introduction into the American market in 1975, the new innovative engine met United States emission standards while maintaining top-notch fuel economy. Honda consequently captured a sizeable portion of the United States market.

This entire process involved only five or six years, but it was time enough to allow an innovative solution to two desired social goals—low emissions and excellent fuel economy.

THE MEANS OF COMPLIANCE AND NOT THE ENDS

During the last ten years, a number of social laws and regulations have had a tremendous impact on our approach to technological development.

Handled properly, such laws could have spurred new efforts to seek solutions to problems. Competition to find solutions would have been stimulated. This, however, has not been the case. Regulation in large part has not been goal-oriented. Rather than requiring a specific output, such as worker accidents per million hours worked or parts per million of a substance in a plant effluent, the regulations have proceeded to tell the regulated how to achieve the result. The following example serves to illustrate this point:

The Consumer Products Safety Commission is attempting to dictate product design by proposing a ban on certain sizes of aluminum alloy wire. Their proposal specifies only copper rather than establishing performance standards for all materials. The latter course would leave industry innovative flexibility to develop alternate *safe* ways to accomplish standards with the best and lowest cost materials that can be developed. Copper may not be available at a reasonable price to meet future needs. Alternate materials such as aluminum, aluminum alloys, sodium, and glass filaments need to be further developed. The point is that the proper regulatory posture would be for the agency or agencies involved to refrain from directing *how* companies should reach a certain performance standard. Rather, the agencies should hold each company to a performance standard, concentrating on the desired end result, and permit each company to reach a solution through its own innovative efforts.

Where do we go from here? Means must be found to streamline the regulatory process. The process needs to be made more responsive and realistic. If we see regulation as a positive, cooperative venture instead of a punitive, contentious activity, we can be confident that the guidance given will have relevance.

The following revisions in United States regulatory policy—all interrelated—are seen as vital first steps in improving the regulatory climate.

1. *Minimizing inconsistency of regulation.* Each regulatory agency should issue a long-range statement of regulatory intent that could serve as a guideline for both the agency and the regulated industry.

This statement of intent should require appropriate notice prior to any changes to accommodate long-range planning of the regulated party.

When two or more agencies are developing regulations or policy on a single issue or interdependent issues, an interagency coordinating group should be formed to assure consistency.

2. *Reduction in delay in adoption of innovation.* In considering and establiishing regulations and policies, and even in proposing legislation, United States governmental agencies should be required to examine the impact of their actions upon the worldwide competitive posture of

United States industry. The conclusions of this examination should be made public.

3. Reduction of inflexibility of administration. We must insist on a scientific and technical basis for regulation. We must use our technological tools to solve technological problems. We must understand that science has limits, but we should not let this understanding keep us from using the science we do have or from making new scientific discoveries. The agencies should administer the laws as they were written and intended and not extend the laws through regulation. Where we see agencies overstepping their statutory authority, we must seek appropriate remedy by cooperative effort. Failing that, the courts must be used decisively.

4. Concentration on ends rather than means. Regulations promulgated to achieve desired social goals should be limited to standards of performance. They should not dictate the processes used by industry to achieve the standard. Such a refocusing of regulations on goals would foster innovation both in meeting the standards and in allowing industry to devote more resources to product and process innovation.

In view of the complexity of the regulatory problem as it impacts science and technology policy, we have our work cut out for us.

Solutions of environmental problems, for example, are independent of the economic system, although many critics think and act otherwise. But in this country by and large industry holds the key.

• Industry must continue to be viable, to make profits. Without profits, the government cannot collect the taxes that fuel the social engine.

• Industry must make profits, but do so in within a framework of social responsibility unknown in earlier times.

• Only industry is considered to be self-serving by regulators, legislators, activists, and the news media. This is because of the presumed dedication of management to its constituency. Therefore, an awesome credibility gap exists that must be overcome. Overcoming this credibility gap requires understanding by both management and regulators. Industry has most of the technology and some of the other resources to help make a more cooperative approach feasible.

I suspect that in the face of growing public confusion and government overreaction, some of us have become too defensive—and so worried about reacting that we do not initiate actions of our own. The challenges are many. Here are some to consider and discuss.

1. All who have resources must become involved in the regulatory process as early as possible. Our involvement should be with constructive suggestions and not just criticism of proposals.

Douglas Costle, administrator of the Environmental Protection Agency, has challenged business to get off the soapbox and get around the conference table. Although I'm not ready to give up my soapbox, I think the challenge is a fair one. As Mr. Costle says, "It's time to recognize that the American people are committed to regulatory objectives, but that they want us to achieve them at the lowest possible costs."

I believe that the academic community should have a role in the regulatory process. When government and industry are adversaries, they form a two-legged stool. With academia as an active participant, and with a positive, cooperative approach, we might be able to build a three-legged stool which would have solidarity and, most importantly, credibility.

2. Means must be found to deal with insignificant risks in rule-writing. Similarly, means must be developed to deal with benefits in acceptable ways in the risk/benefit controversy.

3. Means must be found to correct bad legislation—legislation calling for zero anything and best available anything. Industry has had its deficiencies in the past and these must be recognized. Regulations should aim to minimize unsafe conditions and to dispel ignorance about work practices.

4. Means must be found to help the media explain complex issues simply and to minimize the impact of adversary rhetoric and sensationalism.

5. Means must be found to move the regulator out of policy-making roles. This is the position they assume when Congress lets the regulators write the rules.

6. Means must be found to require industry, academia, advocacy groups, and regulating agencies to prove their respective points with data. Perhaps independent "laboratories" could be established for studying issues of public concern.

7. Regulation in the past has been too much of a knee-jerk response and has often been based on bad science.

8. For business contracts to be successful, all parties must benefit. I believe that regulation can be thought of in the same light. Good regulatory policy should be able to benefit all of the parties—society, workers, academia, and management.

Overview of Policy Issues:
Panel Report

Panel Chair: WILLIAM R. DILL
Panel Members: Reynald Bonmati, Lee L. Davenport,
Kenneth Gordon, Richard F. Hill, Paul F. Hopper,
Grace Ostenso, A. E. Pannenborg, and John M. Rozett
Rapporteur: Richard Langlois

REGULATION is a pervasive influence on the economy, technology, and society. And, as Monte Throdahl suggests, it is creating many frustrations for both business and society. Regulatory efforts in such fields as pollution control, energy, product safety, and employee welfare are fragmented and frequently inconsistent among agencies and levels of government. Regulation is often unpredictable, with frequent changes in standards, directions of emphasis, and methods of enforcement. In many cases, government has apparently concentrated on means rather than ends, without specifying goals to be achieved. It has set narrow and sometimes irrelevant rules about details of technology or management practice. We are losing a sense of balance about benefits to be derived from the costs of regulation, frequently by seeking results that are unrealistic in terms of expenditures required or expectations about elimination of risk. Too often regulations take the form of using a sledgehammer to kill a gnat. Priorities and proportionate weighing of benefits and costs need to be brought into better perspective.

Misguided regulatory efforts are seen increasingly to be a drag on the economy. Overregulation has been shown to reduce real economic growth and to contribute to inflation. By directing new investment into relatively unfruitful areas and by adding to the uncertainty of planning, regulation may also be reducing the rate of investment for innovation. It has probably contributed to the slowdown in growth of productivity in American enterprise. Too much of new investment currently goes toward meeting regulatory restrictions, not toward improving products, services, and productivity for the economy.

William R. Dill is Dean of the Faculty of Business Administration, Dean of the Graduate School of Business Administration, and Professor of Management, New York University.

0077-8923/79/0334-0172$01.75/2 © 1979, NYAS

Solutions to these frustrating problems are not easy to find. Consider the problems of inconsistent and uncoordinated regulatory policy, an issue that many scientists, corporate executives, and university professors are now addressing. The only thing worse than the existing confusion and inconsistency would be the pattern of government that would inevitably emerge if clarity and integration were to be achieved across agencies and levels of government. Confusion and inconsistency are inevitable consequences of pluralism and decentralization in government; the alternative of more concentrated and powerful central Federal authority would be a threat to the freedoms that are important to us all.

Not all of the inconsistency, further, is unwelcome from industry's point of view. Much of it results, in fact, from companies behaving like the "single issue" advocates whom they deplore and lobbying for localized and tailored exceptions to general regulatory directions that suit their interests. Pluralistic regulation is not simply a result of sloppy government planning. It is a result of the kind of society in which we live.

Nevertheless, regulators, those being regulated, and those whom regulation is meant to serve have an obligation to explore the more serious cases of interagency and interlevel conflict. Where common goals do exist and where common procedures and standards would work, savings should be explored. The possible savings are not only economic in nature. They include intangibles such as improvements in attitude that can result on both sides when antagonism over contradictory and overlapping approaches and jurisdictions is removed.

Some frustration over regulation in the United States seems to result from the feeling that there is less antagonism between business and government in many countries overseas. Perhaps so, and perhaps not. Government has played a stronger hand for a longer time in Europe and Japan, both in managing the economy and in developing technological partnerships with industry. Businesses there may not be subjected to more acceptable forms of regulation, but they may rather simply be more used to living with it. The one major difference—and it is a very important one—is that in most other countries international trade represents a larger share of economic activity. Government-business collaboration to help nations compete on economic terms and to assure strong export positions has long been a reality throughout Europe and Japan. It is just now being recognized as a need for the United States.

Some of the pressures for regulation can be viewed as the public's response to the growing size and power of corporations and to the longer time spans over which they plan, develop technologies, and make investments. Customers, investors, and citizens want a voice in the commitments a company makes that might affect health, safety, or other amenities of life before investments are defined and before a course of action, which may prove damaging, has been undertaken. The public is no longer content simply to wait and judge corporate performance by results, in terms of decisions to buy or not to buy a product or in terms of deciding whether to hold or to divest stock. They want assurance and participation in making sure that long-range investments will be sensible ones.

Bringing the public into early stages of strategic planning is a disturbing and threatening idea to business because it violates important traditions of privacy and competitive secrecy. It does not assure that greater expertise will be brought to bear in making decisions, and it sometimes has resulted (as in the case of many environmental reviews) in long and costly delays in making decisions and starting investments.

Yet experiments in external consultation are being expanded and may well become a way of life for the future. Such a review helped greatly to lay the groundwork for expansion of coal-mining programs a few years ago. Monsanto consulted widely with outside experts and advocates to make sure they had anticipated as much as they could in planning for introduction of the acrylonitrile soft-drink bottle. The food industry has reached out to government, universities, and consumer groups to set up a Food Safety Council as a new forum for trying to foresee ways to improve definition of regulatory standards and industry practice in evaluating new food products.

Universities must join forces with industry and government to help address issues of improving regulatory objectives and practices. There are many areas in which research is needed, and in which properly led academic research will carry more credibility across various sectors of society than the best of industry or government efforts can. Some of this research is on basic questions of science and technology, but with more emphasis than heretofore on looking at dimensions of the problem that have regulatory overtones. Research in chemistry, for example, should not only examine the efficacy of new processes for use in manufacturing, but should also look more closely at possible side effects that would affect health or safety. Much more work needs to be interdisciplinary in flavor, and research scientists must become as well acquainted with social questions and issues surrounding their work as production managers and marketing executives have had to become.

Research is also needed on how to regulate. We need to know more so that we can teach more effectively how to make regulatory measurements; how to assess costs, benefits, and risks; and how to provide meaningful kinds of peer review of the kinds of scientific work that are used in regulatory standard-setting.

Beyond research, universities have a major teaching obligation. Academics may not too often be single-issue advocates, but they are too often devotees of a single methodology. Regulatory policy-setting is handicapped because those who participate—lawyers, managers, politicians, and scientists—come from four very separate academic and philosophical traditions. Most are very poorly educated in the perspective and approaches of the other groups with whom they deal. In undergraduate education, in graduate and professional schools, and in postgraduate education for experienced managers and professionals, bridging work is needed, in the form, for example, of courses and seminars that help lawyers to understand engineers, engineers to understand economists, and economists to understand politicians. New York University is trying at present to build such an all-encompassing Center for Studies of Regulatory Policy, and it is hoped that many other institutions will do the same.

As a final note, the improvement of regulation does not start with complaints about rules that are already established. Refinements and amendments of these rules are helpful, but the important direction is in anticipatory work—in basic research and teaching, in collaborative planning efforts such as those undertaken by the Food Safety Council, and in briefings and colloquia with groups that do the most drafting of regulatory rules and procedures. In particular, emphasis is placed on programs involving Congressional and state legislative staffs and the staffs of major continuing agencies who are central to the drafting process. Early meetings, discussions, and educational programs can be nonadversarial in tone. They can yield both better laws and better collaborative understanding for administration of the law.

The Social Responsibility
of Scientists

DOROTHY NELKIN

MORAL PHILOSOPHERS have long argued that children and madmen are the only beings not fully responsible for their actions: "For as madmen are thought to lack freedom of choice, so children do not yet possess the power of reason in a developed form." Clearly scientists neither lack freedom of choice nor are they devoid of reason. Why then would a special American Association for the Advancement of Science committee on social responsibility find it necessary to argue that scientists should assess the potential social consequences of their work? Why then has so much discussion, and indeed often acrimonious debate, revolved around the concept of social responsibility among scientists? I would like briefly to sort out some different concepts of responsibility in science, and then place the discussion in historical perspective in an effort to clarify the ambivalence about the implementation of responsibility in the present social context of science.

CHANGING CONCEPTS OF RESPONSIBILITY

The concept of responsibility in science has several dimensions. First, responsibility is often defined in terms of the methodologic obligations of scientists in carrying out their research. June Goodfield caricatured this methodologic ethic as follows:

- Thou shalt not cheat.
- Thou shalt make thy experimental results available.
- Thou shalt not covet thy neighbor's x-rays or diagrams.
- Or, if thou dost covet them, thou shalt not acquire them.
- Or, if thou dost acquire them, thou shalt have the proper decency to keep quiet about it.

Second, the notion of responsibility often refers to the obligations of scientists to those involved or immediately affected by research procedures. This includes human subjects, informants, and the technicians or

---

Dorothy Nelkin is a Professor in the Program on Science, Technology and Society and the Department of Sociology at Cornell University.

0077-8923/79/0334-0176$01.75/2 © 1979, NYAS

neighbors subjected to immediate risks. Whereas in the first instance norms define the responsibility of scientists, in the second formal guidelines leave relatively little discretion.

Third, responsibility implies sensitivity to the political and economic context of research and, in particular, to the effect of the sources of funding on the quality, integrity, and utilization of research. For example, bacteriologic research funded by the Department of Defense will have different applications from that funded by the National Institute of Health.

Finally, the most difficult and nebulous dimension of social responsibility is the obligation of scientists to use their special expertise to call public attention to the dangers of new technologies and, indeed, to the potential use and consequences of their own research. This aspect of responsibility, with its implications of political activism and public dispute, is the greatest source of ambivalence; for political activism also carries implications for the autonomy of science from external control.

That scientists have a responsibility for the use and consequences of their work is certainly not a new concept. In the early part of the century, geologists, agricultural scientists, biologists and physicists systematically tried to bring their scientific knowledge to bear on improving public policy. After World War II, the movement for social responsibility took a new and more active turn with the efforts of the atomic scientists to challenge public policy concerning military technology. Deeply disturbed about dangerous military applications, these activists voiced their dissent in journals such as the *Bulletin of the Atomic Scientists*, and they formed new organizations such as the Federation of American Scientists and the Scientists' Institute for Public Information. Using the press and the podium, they informed the public about the implications of atomic energy, hoping thereby to influence government policy.

By the late 1960s, the postwar ethic of social responsibility reached a still more active phase as some scientists organized to oppose the antiballistic missile, and others to protest military research in universities. In the political context of that time, a new critical science movement emerged, and this has changed the character of scientific activism in several important ways. Except for the postwar concern about military applications of science, social responsibility in most policy areas had been defined in terms of contributing scientific knowledge to public affairs. Since the late 1960s it has rather implied the challenging of public policy in a wide range of areas. Political relationships have evolved from "prudential acquiescence" to overt dissent as activists in the scientific community have extended their political interests beyond military issues to the en-

vironmental and antinuclear movements and to questions about biomedical and genetic research. They have critically examined research
practices, analyzing the possible risks and abuses of science, and ques-
tioning whether some research should be done at all. The recombinant
DNA debate, the XYY dispute, the genetics/IQ controversy, the sociobiology debate—all embroiled scientists in public controversy as they
questioned the risks and consequences of research.

The most striking change in the character of scientific activism is its
political orientation. The 1940s activists were motivated by personal and
moral concerns; today's activists have developed a political perspective.
They are far more cynical about the responsibility of scientists than were
their ancestors in the 1940s, who had sought above all to preserve the autonomy and self-regulation of science. Indeed, today's critics seek guidelines and formal controls to assure the integrity of research and protection of the affected public.

THE ORGANIZATION AND TACTICS OF SCIENCE ACTIVISM

The implementation of social responsibility has also taken on new organizational dimensions. The postwar scientists' movement was for the
most part composed of individuals who felt a personal sense of responsibility. Political action was relatively unorganized. Today the scientists'
movement has a more complex political base; it includes the so-called
"critical science" or "public-interest science" groups as well as traditional
professional and scientific societies.

Scientists with quite different political orientations identify with this
movement. Some are highly critical of established science and technology, seeking basic social and political change. The most visible participants in the recent scientists' movement are those essentially "professional activists" who speak out on very diverse science and technology
policy issues. While these people are few in number, they count on the
support of a much larger group of young scientists who were politicized
during the Vietnam War and find these issues an outlet for their political
energies.

But most scientists take a more pragmatic and conservative approach,
implementing their sense of responsibility by providing information and
technical assistance to citizen groups on specific issues. Convinced of the
political efficacy of education and information, these public-interest
scientists are, ideologically, direct descendents of the post-World War II
activists, often using the same media (for example, the *Bulletin of the
Atomic Scientists*) to document their position. However, while scientists

in the 1940s went to the public arena as a "reluctant lobby," as if such action violated the norms of science, today's activists engage in the political forum with verve and enthusiasm.

Public-interest scientists have begun to institutionalize the concept of social responsibility: careers now exist for scientists in public advisory roles, networks of scientists are available to advise citizen groups on particular problems, and even a division of the National Science Foundation supports scientists who wish to provide their expertise to citizens. Organizations of politically active scientists in diverse technical areas have proliferated: The Center for Scientists in the Public Interest, the Union of Concerned Scientists, the Clearing House for Professional Responsibility, Science for the People, Scientific Workers for Social Action, Aerospace, Computer Professionals for Peace, and the National Coalition for Responsible Genetic Research, just to name a few.

The trend is not simply American: critical scientists in Great Britain are organized in Health Hazards Advisory Groups and in the British Society for Social Responsibility in Science. In France, GSIEN (*Groupe Scientifique de l'Energie Nucleaire*) is the equivalent of our Union of Concerned Scientists, and in Holland the *Wetenschapswinkels* or science shops are well institutionalized as science advisory units for community groups seeking to challenge technological policy.

Tactically, the most striking feature of the new scientific activism is its public nature, and the willingness of scientists to engage in and indeed to abet political controversy. Clearly scientists often disagree, but their disputes are contained within the scientific community with its well-established procedures for collegial review. However, those active in current controversies are rather inclined to use a political forum—the popular press, public testimony, or litigation. They seek a broad constituency, appealing to groups external to science to support and to implement their concerns: For example, Cambridge activists who opposed Harvard University's plans to build a recombinant DNA laboratory ran public workshops and talked to technicians and local community groups. They presented their concerns to Mayor Vellucci, known for his willingness to attack the academic community.

To be sure, the atomic scientists' movement of the 1940s also lobbied, lectured, contacted the press, and wrote popular articles. But they were seeking a public constituency in order to support their concerns about military policy. In contrast, today's activists seek a public constituency in order to increase the accountability of science. While activists in the 1940s fought to isolate research from political control, their recent counterparts want to increase political interaction. "Whistle-blowing," "demystification," "accountability," "participation"—this is the rhetoric of social responsibility in the 1970s.

Traditional scientific societies, faced with constituent pressure, have inevitably reacted. The American Chemical Society produced public documents on "Chemistry and the Environment"; the American Physical Society created a forum to discuss policy issues in which science plays an important role. Professional associations have groped for ways to protect their members who "blow the whistle" on harmful scientific or industrial practices, to develop codes of ethics that would constrain potentially dangerous research, and to extend the norms of responsibility to include consideration of the social implications of science.

The Asilomar conference of 1975 presented a striking example of an organized professional response to pressures for social responsibility. Convening molecular biologists from all over the world to discuss the biohazards and the problems of experimental safety in recombinant DNA research, Asilomar was called a "model experiment in social responsibility" in which scientists were seen trying to do what they felt was right for the public, rather than for themselves. Asilomar, however, provoked a far broader debate than scientists had anticipated. In many ways recombinant DNA became the "atomic bomb" of the 1970s—a symbol of concerns about science and its consequences. As the public became increasingly engaged in this debate, many scientists regretted their initial altruism. Indeed, Asilomar and its aftermath suggest the ambivalence and fear about the political implications of social responsibility.

RESPONSIBILITY OR IRRESPONSIBILITY?

Scientists who extend their skills to the political arena have always faced the sanctions of their colleagues. Implementing social responsibility implies political activity in a subculture that has long assumed such activity to be destructive of scientific endeavor. Prior to World War II, Bernal observed that "any attempt on the part of the scientist to think for himself outside his own field exposes him to severe sanctions. . . . It is argued that in the interests of science it would be far better for him not to do so." In 1978, the president of a professional society commented, "Science is great because it aggrandizes man. Politics is tawdry because it belittles."

Clearly the political activity of some scientific leaders is perceived as useful for science. But those who challenge existing practices in controversial policy areas—and that is, after all, the essence of social responsibility today—are regarded with ambivalence. Informing the public about the potentially harmful consequences of science and technology necessarily engages scientists in public debate. Such political involvement and open indication of disagreement among scientists can destroy the image of neutrality that has long justified the autonomy of science.

Concerned about external control, many scientists regard the implementation of social responsibility as scientifically irresponsible. They raise a number of questions: Is it responsible for scientists to publicize their beliefs about the toxicity of a new chemical or the adequacy of an environmental standard before their findings are submitted to professional review? Is it responsible to warn of the potential consequences of research when these are difficult, if not impossible, to anticipate? Given the difficulty of formulating a systematic notion of the general welfare or the public interest, how can one define the notion of responsibility and establish appropriate standards? Does trading on the public image of science or on a scientist's own reputation to influence public policy represent responsibility or merely the exercise of power and influence? And how can the exercise of social responsibility be distinguished from efforts to promulgate particular values?

In sum, scientists (neither madmen nor children) remain ambivalent about the concept of social responsibility. This ambivalence, based on concerns about increasing external control over science, reflects the basic contradiction between the cognitive and pragmatic dimensions of science. Is science the pursuit of truth or the pursuit of useful knowledge? Is science to be defined primarily in terms of a carefully disciplined process or as a professional activity?

This contradiction poses increasing dilemmas as scientists extend the scope of their professional activities. Science today is a social, political, and economic resource, used to formulate and legitimize major policy choices. Scientists today serve as advisors to policy-makers, consultants to private enterprise, expert witnesses to courts, technical administrators, social critics, popularizers, advocates for community groups, and workers in large technical bureaucracies. In the present stage of the development of science and its relation to society, these multiple roles must be considered as intrinsic to the scientific enterprise. But they also introduce institutional demands that may conflict with the value system of science as a cognitive activity. The autonomy of science is viewed as fundamental to healthy and productive research, but it is in fact limited in most professional settings. Those employed in industrial settings are not free to raise critical questions if calling attention to scientific lapses would jeopardize their personal security. Scientists who work in a context of secrecy based on the proprietary concerns of industrial firms are constrained in the sharing of information. Dependent on external fund-

Overview of Policy Issues:
Panel Report

Panel Chair: D. BRUCE MERRIFIELD
Panel Members: Laurence Berlowitz, Sidney Borowitz,
William D. Carey, Philip Handler, Milton B. Hollander,
John Holmfeld, Gösta Lagermalm, Steven Marcus,
Thomas McCarthy, Richard B. Opsahl,
Vivien B. Shelanski, Allen M. Shinn, Lowell W. Steele,
David Swan, and John Thompson
Rapporteur: John Lightstone

THE CONCEPT THAT INDIVIDUAL scientists and engineers
have a responsibility to society that goes beyond the integrity of their
own work began to emerge first with the advent of the atomic bomb but
has expanded rapidly in scope over the last decade. In a sense, this is a
symptom of a more subtle process which perhaps also explains the
remarkable ascendancy of the "industrial democracies" in modern times.
It is indeed a remarkable fact that about 10 percent of the world's population living in the industrial democracies has achieved a quality of life that
has far surpassed that of other cultures, many of which have existed for
thousands of years.

The Frenchman M. Revel recognized this many years ago, and pointed
out that the opportunity for self-actualization in the industrial democracies has led to the emergence of two factors new to the world. One of
these is the accumulation of a "critical mass" of educated people capable
of communicating on a conceptual level. The second is the advent of instant communication (radio, television, and so forth) which allows very
rapid interaction among them on any issue. Revel predicted that the
combination would be explosive and that the resulting multiple forms of
discussion and debate at many levels could change the world by the end
of the twentieth century and create a degree of ethical understanding and
sensitivity never seen before.

A simplistic further breakdown of this "stochastic"* process of communication identifies four sequential steps:

* A stochastic process is a multiple-step, multiple-interaction time-based process, which
is sufficiently complex that direct cause and effect may not be discernible, but movement
over a period of time does occur.

D. Bruce Merrifield is Vice-President of Technology, The Continental Group.

0077-8923/79/0334-0183$01.75/2 © 1979, NYAS

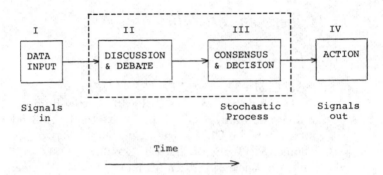

The logic of the sequence is that concerted action is not possible without consensus, and consensus requires extensive discussion, debate, and ventilation of opinion. Finally, effective discussion and debate can only occur when a common data base or set of facts is available as a point of departure for further input of ideas and feelings.

Basically it is this sort of process that is both at the heart of the participative goal-setting and goal-implementation of the industrial democracy and which also describes the vigorous "social responsibility" debate that has emerged in the last several decades.

Dorothy Nelkin, in a paper in this volume, articulated the stages of development and some of the traumas that have been involved in the stochastic process responsible for the awakening of the scientific community. The perspective of scientists, she points out, has evolved from a methodologic ethic[†]:

- Thou shalt not cheat.
- Thou shalt make thy experimental results available.
- Thou shalt not covet thy neighbor's x-rays or diagrams.
- Or, if thou dost covet them thou shalt not acquire them.
- Or, if thou dost acquire them, thou shalt have the proper decency to keep quiet about it.

The term of this evolution is a sense of concern: first for those involved in, or immediately affected by, research procedures (hoods, goggles, safety shields); then for the political and economic context of research (particularly that the source of funding should not affect the quality or integrity of the research); and finally, for the potential dangers and/or "misuse" or research results (the nuclear arms race, pollution of the biosphere, and so on).

The dangers of involving an uninformed public in debate on technical issues (recombinant DNA, for example) with the possibilities of hys-

---

[†] Quoted from June Goodfield.

terical overreaction and undesirable regulatory legislation emphasize the importance of establishing an effective (first stage) data base as an essential prerequisite to an effective (second stage) debate and discussion. The second stage is always the most emotional and uncomfortable stage, and it is important not to misunderstand the slings and barbs or even occasional acts of violence which can accompany this stage. They are just symptoms of the "vigor of the process."

Dorothy Nelkin traced the early triggers of the awakening process which first involved atomic scientists concerned about military technology, the hydrogen bomb, and later the intercontinental ballistic missile. In the late 1960s, however, the debate began to extend beyond military issues to environmental, biomedical and genetic research areas. The genetics/I.Q. controversy, the XYY dispute, and the sociobiology debate all raised questions involving risk/benefit analysis and became heavily politicized. Moreover, science activism itself became institutionalized, and although the activists themselves were few in number they were supported by a much larger group of younger scientists politicized during the Vietnam war. A host of organizations now exist worldwide which are solely dedicated to the task of "whistle-blowing, demystification, accountability," and so on—to quote the rhetoric of the 1970s.

One issue raised by Professor Nelkin is the degree to which this ongoing debate is responsible or irresponsible. Uninformed public reaction can lead to undesired controls in the eyes of the scientific community together with a dilution of the image of objectivity and neutrality that has long justified a certain autonomy for scientists. Also, in some degree, premature public debate of new findings may subvert the peer-review process which is basic to scientific inquiry. Is science the pursuit of truth or only the pursuit of useful knowledge? These are unresolved dilemmas which extend the scope of science beyond its older dimensions. Science is now a social, political, and economic resource vital to survival and to the quality of life.

But institutional demands will continue to conflict with the values of science as a cognitive activity. The degree of autonomy required for productive scientific research will always conflict with the possibility of unanticipated consequences that might result from new discoveries. Society is and will continue to be in the throes of a vigorous renegotiation of its relationship with science. This will be a continuous stochastic process of change accompanied by both risks and benefits.

Nevertheless, it is clear that a better understanding of "good process" is important to an era of continual renegotiation. A good data base effectively disseminated to all interested publics is a prerequisite for effective debate. Pluralism and tension (the emotional interaction of diverse opinions and points of view) are essential for a productive discussion/debate stage. Consensus cannot really occur without these two steps, and concerted action is impossible without consensus. A cooperative role (with effective incentives for cooperation) involving government, academia, and industry is an essential part of the pluralism needed to distill the greater wisdom of the society. And an effective staging of the stochastic process can accelerate, smooth, and make more productive the way to consensus and concerted action. It is indeed a positive process and one hopes that M. Revel is correct in his prediction of its positive consequences. Certainly we live in a moment of rapid change that is unique in the history of civilization.

Foreign Policy*

RODNEY W. NICHOLS

INTRODUCTION

VIRTUALLY EVERY MAJOR scientific and technological policy issue comes within the span of *both* domestic and foreign policy. Unquestionably, all the "functional" topics—such as Food, Health and Energy that are considered elsewhere—have to be thought about today in global terms.

Even worse, we must keep other themes in mind. For example, two subjects are not explicitly included in this conference's program: Defense and Space. It would have been inconceivable to have an international meeting on science and technology policy in the 1950s or 1960s without devoting considerable attention to national and international security. Similarly, other important foreign-policy-related topics were left off the formal agenda for reasons that no doubt related to limits of time; e.g., the problem of population growth, and the concern about proliferation of nuclear weapons.

Furthermore, several "cross-cutting" subjects are not explicitly on the program here because they are less important for industrialized nations. But because some of these subjects are crucial for developing countries, they create foreign policy problems for developed nations. In this category would be issues such as the priorities for industrialization; the needs for much broader education and training; and the problems in retrieving and distilling information (at the working level in R&D and at the policy-making levels) for national priority-setting affecting R&D.

So while our assigned task is difficult, it would have been even more complex if the full agenda were known. Bringing coherence to this domain is long overdue. A brief outline may help to structure discussion and reveal the envelope within which most of the current policy debates take place (see TABLE 1).

---

* The author spoke from notes. This paper was prepared later and, although it is somewhat longer than the 20-minute presentation at the meeting, it retains the aim of opening up many subjects rather than assessing a smaller number in greater depth.

Rodney W. Nichols is Executive Vice-President of The Rockefeller University.

0077-8923/79/0334-0187$01.75/2 © 1979, NYAS

TABLE 1

---

1. Goals: How to Define the Subjects?
   - Power and welfare among "the worlds"
   - Deterrence and defense
   - International security and peacekeeping
   - East-West issues
   - North-South issues
   - Global concerns and principles

   - Historical trends
   - Future stresses
2. Policy Puzzles: The Urgent, the Important, and the Uncertain
   - Competing geopolitical views
   - Economic models, policies, consequences
   - Priorities for resource allocation
   - Realistic time-tables for results
   - Evaluation and output indicators

   - Cross-cuts on every S&T policy
   - Infrastructure of ideas
3. United States Patterns: Turning a Corner or Turning Away?
   - Congress
   - Executive branch
   - Industry and organized labor
   - Universities

   - Media, public understanding, influential ideas
4. International Patterns: Centripetal or Centrifugal Forces?
   - Bilateral arrangements
   - Alliances, regional systems, coalitions
   - UN system
5. Illustrative 5-Year Priorities: What is S&T in (for) "Foreign" Policy?
   - Traditional missions of national governments
   - New missions linking industrialized nations
   - Economic competition
   - Cooperation with LDCs
   - New initiatives for policy-analysis

GOALS: HOW TO DEFINE THE SUBJECTS?

The first heading states the main question I have already introduced: In considering the goals of foreign policy, how can we bound the many global subjects and complex international programs that depend in part upon science and technology?

For example, goals related to *power* are still significant for most officials responsible for foreign policy. An astute French observer commented some years ago that "neither legions nor raw material nor capital are any longer the signs of instruments of power. Force today is the capacity to invent, that is—research; force is the capacity for converting inventions into products, that is—technology." In this observation lies the kernel of what many nations see as a critical role for science and technology serving foreign policy's purposes.

Consider the newspaper's categorization of the globe into various "worlds"—the "first world" of the OECD industrialized nations, the "second world" of the Soviet Union and its Eastern European allies, and the "third world" which is the developing countries. (More discriminating journalists divide the globe further into fourth and fifth worlds, based roughly upon resources per capita.) These numbers suggest a power-related rank-ordering. In fact, we often measure changes in international relations, and the status of a nation's geopolitical power today, in terms of the degree to which nations have applied technology successfully to building up their domestic and foreign strengths. Will national power in the future continue to be proportional to the capacity for, and the productivity of, research and development?

A related major goal derives from the hopes of people everywhere for increased *welfare*—or, in the current phrase, greater "equity." For the developing nations, the aim is to draw on the world's technological experience to increase economic activity and meet the basic human needs of their people. For the industrialized nations, a meaningful policy to improve welfare increasingly demands better *control* of the new technology that is required to achieve humane purposes.

With this admittedly superficial skimming of the profound issues of power and welfare in the world—and the way in which science and technology serve those goals—we can race rapidly through a number of other crucial goals of foreign policy.

Deterrence and defense are of interest to all nations. For the OECD nations, we think of the NATO alliance and the Strategic Arms Limitation Talks as prime examples of the use of advanced technology to serve the goals of war-prevention and arms control. Indeed, it is a cliché to note that without the most sophisticated technology, there could not even be hope for assured verification (by national means) of the proposed SALT II treaty.

Turning to a likely superpower of the next century, the People's Republic of China includes both defense and science/technology among its "four modernizations." China says that science and technology will be essential to assure success in its defense and, thus, in its foreign policy. However, its alliances may shift over the next generation as international technological trends unfold.

As a final example, in almost any scenario about how the superpowers

would avoid World War III—or limit escalation and possible consequences, should a major war seem to be threatened—the most advanced
technology would be employed by the highest officials in communications, command, and control.

International security and peacekeeping have only begun to be studied
in the robust analytic and institutional traditions that have been characteristic of past research and development on many other subjects. Yet the
shrewd use of advanced sensors was important in making possible certain temporary peace agreements in the Middle East. Other advanced
technologies—in transport and communications, for example—may
keep small conflicts from flaring into larger fights by allowing small
forces to contain larger ones for a short period while diplomats confer.
Yet what shall we do about reducing the arms trade in conventional weapons, about controlling terrorism, and about dampening regional arms
races? Could any technologies be helpful in new ways of *solving* these
problems? Even if there were no technological "fixes" for such largely
political issues, the non-nuclear technologies of "conventional arms"
turn out to be very important for diplomats to master; and the Foreign
Service pays little attention to this (or any other) technical trend.

In the category of *East-West issues* are compressed many topics that
relate to technology. For instance, debates about controls on the exports
of certain products usually involve the possible long-range military and
economic impacts of *multiple* uses of the technology embodied in equipment that would be exported initially for one narrow purpose. The difficulties in dealing with classified technical information—and particularly engineering data, rather than general scientific materials—pose major
problems for foreign policy debates on such exports. Another example of
how technology relates to East-West issues is the ability of major (and
lesser) powers to employ modern airlift capabilities to transport troops
across large distances and thus project the image of power as well as the
actual instruments of force. Historically deep-running strategic stakes
emerge at the interface between international technological trends and
East-West balances of military and economic forces.

Similarly, *North-South issues* include an enormous range of topics
that almost always involve technology. Rarely, however, is technology
the most important variable. For example, the clamor for a "Code of
Conduct" governing multinational companies (MNCs) often encompasses demands for technology, but it is mainly the sheer economic
power of the MNCs that has produced the perceived need for such a
Code. Patent rights are debated internationally, because many LDCs
believe they have a "right" to essentially any technology anywhere.

More broadly, from the perspective of the LDCs technological "dependency" upon developed countries, every contemporary dialogue about North-South relations touches on how to increase the level of international research and development that relates directly to the needs of developing countries.

Many challenging, long-run topics can be placed under the heading of *global concerns and principles*. Perhaps the most dramatic change in the world during the 1970s—a change that has put science and technology more squarely in the middle of the diplomatic map—has been the sharply broadened understanding that *global* problems really matter. Consider, for example, the new senses of interdependence in relation to earthquake prediction, changes in climate, incidence of famine, supplies of energy, surveys of resources. Even public health is viewed internationally in new ways, although health for a much longer time had been understood as a worldwide responsibility owing to the risks of epidemic infectious diseases. Cogent subcategorizations can and must be made among these global concerns as they affect specific countries. But technology is drawn from world-wide sources and partly for this reason diplomats around the world have been forced to deal more frequently with scientists. Furthermore, the universal goals of human rights resonate with the international ethos of the scientific community.

Our discussion-outline shows a dotted line at this point offering two further headings about themes of a different kind: historical trends, and future stresses.

Mr. Delapalme presented interesting data regarding the recent industrial production trends in Europe, the past dramatic changes in demographic projections, and the less dramatic but nonetheless decisive changes in power relations among the major industrial nations within the past 100 years. All of these data underscore a remark that Brzezinski made a few months ago: "There is a redistribution of both political and economic power in the world today . . . this means that the older industrial countries have to rely increasingly on technological innovation to maintain their place in the world."

Just as Brzezinski was looking historically in that remark, we must refine our analysis of how and why past trends have driven so many issues of science and technology into the concerns of foreign policy. We must also be clear about the likely future stresses—such as on food supplies, on energy resources, and on our organizational mechanisms for international collaboration and conflict-resolution.

I have tried so far to illustrate—briefly, yet systematically—that the perspective of foreign policy complicates many issues in national science and technology policy. Let us now turn to some specific "policy puzzles" that are raised in trying to resolve international issues.

POLICY PUZZLES: THE URGENT, THE IMPORTANT, AND THE UNCERTAIN

In this second part of the discussion, we will follow the same pattern as in the first part. Let us categorize major areas of difficulty.

To begin with, *competing geopolitical views* guide the directions of science and technology policy. For example, if the intentions of the Soviet Union were known accurately, that knowledge would certainly affect the way in which the United States proceeded in defense R&R; in the absence of reliable knowledge about USSR goals, competing estimates arise and defense R&D can be viewed as insurance. Similarly, longer range projections for Asia often turn upon estimates of the technological strength of Japan and of China, in relation to the U.S. and the Soviet Union (as well as to other smaller, industrializing countries in the region). More generally, if the United States saw technology as its principal lever for ensuring both military and economic power, then our political uses of such technologically based power would depend heavily upon making larger, more visible investments in new technological capabilities. To the extent that any country wishes to use technology for pursuing foreign policy goals, it must take account of long term political trends reflected in national technical efforts around the world. Whether or not one takes classical geopolitics as the dominant perspective, technological capabilities probably will play a key role in geopolitical trends.

A second area of puzzles, particularly pressing at the moment, is in the use of *economic models* to test the consequences of alternate policies. To oversimplify somewhat, there is no "theory of R&D" that holds water in our present economic models. In DCs, we cannot compute how to achieve the right level of "innovation"—the level sufficient to avoid stagflation and sustain increasing productivity. For LDCs, we cannot demonstrate convincingly what will be the varied immediate impacts of technical change during the process of modernization; and even less well can we predict the more long-range cultural changes caused by "development." Thus, for two of the most far-reaching economic frustrations in the world today, the apparently critical role of technology remains murky.

Stunning evidence about these crippling gaps in our economic knowledge emerged last year from a *Fortune* poll of professors of economics at 55 American universities. The results of the poll showed an extraordinary loss of confidence in the ability to make accurate macroeconomic forecasts and deep doubts about the usefulness of any government-stimulated interventions in the economy. In such a situation, it is not surprising that the more complex intereactions of technology with foreign economic policy are wracked with uncertainty.

Along with the global benefits from the introduction of new technology—such as in the agricultural Green Revolution—come undesirable side-effects that, even though exaggerated by some critics, are nonetheless real. Even when there is a major achievement in one country—e.g., Malaysia's successful R&D on natural rubber—other countries have great uncertainty about their technological choices with different natural resources and different social systems. In fact, the fad about "appropriate technologies"—usually meaning comparatively small scale and "less advanced" technologies that are adapted to a particular environment—has been stimulated by the assumption that if technology were more understandable, the consequences of its use would be better anticipated. Yet that is rarely true. In short, the social and economic components of S&T policies are important areas for further research in most international choices about technology.

One of the reasons that economic issues loom so large is the major puzzles concerning *priorities for resource allocation within any R&D establishment*. To begin with, there are trade-offs between what roughly can be categorized as the domestic purposes *vs.* the international purposes for programs of science and technology: e.g., in biomedical research the U.S. must choose how much research to devote to cancer *vs.* tropical diseases. Another trade-off affecting the "third world" concerns (a) investments in general scientific and technical education *vs.* (b) investments in the urgent, more problem-oriented activities such as population control or water resources. None of these trade-offs is easy. Many of them emerge in stark terms during decision-making within national governments and international institutions. There is little analytical guidance for confidently striking such judgements in allocating scarce R&D resources.

Another puzzle concerns the *time-tables for action*. Many developing countries show a remarkable lack of realism about the *long* time-periods required for building scientific and technical institutions. Similarly, many industrialized countries show astonishing apathy about how *urgent* are the South's problems requiring science and technology to meet humane goals in development. Without a world-wide consensus on what the time-tables actually are—or at least a less politically charged climate for reconciling the different views about these time-tables—there is slim hope for sustained action on broader, coordinated efforts.

Finally, among such analytical problems, consider the difficulty in *evaluating results*. We have precious few indicators about the outputs of research and development programs, even the mature programs in advanced countries. We have even fewer reliable indicators about the results of technical assistance efforts abroad (not to speak of the defense area, in which often what is a technological success to one observer is a failure to another observer). Since research and development projects are means to ends—and, more generally, scientific and technological skills are diffused throughout a society—evaluation is extremely hard. Indeed, foreign policy itself is often viewed in terms of a continuing *process* in diplomacy, rather than in terms of the merits of specific end-points of a certain negotiation. Until we have at least somewhat better ways to measure the impacts of the use of science and technology in pursuing foreign policies, there will be little hope for substantially better integration of the policy domains.

This last viewpoint leads to the final sub-topics in this part of the outline (see p. 188). There are *cross-cutting considerations* with respect to every science and technology policy that relates to foreign policy. When we ask about giving technological *aid* abroad, we are faced with trade-offs with respect to its impacts on technologically based *trade* (e.g., steel; electronics; textiles). When we consider East-West relations, we are aware of their impacts on North-South relations in both political and technological terms (e.g., oil prices and supplies; arms trade). When we ask for broader benefits from our science and technology activities within foreign aid, do we emphasize basic human needs or economic growth? If a global perspective showed that domestic economic policies were short-sighted, would R&D policies have a bearing on possible new directions?

In general, we also must contend with subjects for which there is not yet a solid *"infrastructure of ideas"* akin to the structure that, for example, has existed for at least 20 years with respect to debates on national security questions. As implied earlier, historical scholarship and contemporary policy-relevant research must be deepened a good deal more. The few full-time professionals in this field must be supported and a new generation of analysts must be trained so that the ideas and policies are understood as profoundly as the problems merit.

UNITED STATES PATTERNS: TURNING A CORNER OR TURNING AWAY?

We shall shift now to a quick review of the U.S. patterns in dealing with these issues. I will sketch a few points in the spirit of a "national case study," exploring the institutional factors in policy development.

The U.S. Congress has been showing a strikingly broader alertness to the international dimensions of scientific and technological activities. Increasingly, many committees explicitly relate the national to the international R&D scenes. Congress and its supporting agencies have insisted on trying to understand the national impacts of international technological trends, especially, for instance, regarding trade and defense.

In the U.S. Executive Branch, international research and development has been an "orphan." It is quite difficult to obtain even crude data on the scale of science and technology carried out with any international purposes in mind. Recent legislation calls upon the State Department to give much greater attention to the planning, the coordination, and the training aspects of the Department's responsibilities for science and technology serving American diplomacy. In the staffs of the National Security Council, the Council of Economic Advisors, and the Office of Science and Technology Policy, senior staff members focus on every one of the issues mentioned so far. Nonetheless, the various mission-agencies send and receive mixed signals about, and give a generally low priority to, their activities in technology related to foreign policy.

We could summarize all of this in a rather breezy way by noting that the U.S. Government frequently tends to see technology as (a) "a trump card," (b) "a last resort," or (c) "a scarce resource." Such headline-phrases might characterize most of the major U.S. initiatives in recent years concerning R&D with diplomatic implications. For instance, cruise missile technology has been a "trump card" in defense; general exchange agreements on scientific topics have been a "last resort" when other diplomatic efforts have failed or must be nurtured; and advanced computer technologies have been "a scarce resource" to be protected rather than shared. Simplifications with new vocabulary introduce concepts that require definition, which probably ought to be avoided in a subject that is already plagued by ambiguities. But the phrases may help to highlight the occasionally conflicting premises (or purposes) that most governments confront.

It would be impossible to generalize reliably about the roles of industry, organized labor, and universities in the technological relations of the United States abroad. But it is fair to say that an entirely new set of deeper tensions has emerged in the past decade. Competition from around the world has meant that our manufacturers face new problems and our workers face new insecurities. Protectionism is surely not the long term answer. One reconcilation of the many current pressures is to launch, as the Carter Administration has done and as the Ford Administration did earlier, new initiatives in R&D that would help the United

States maintain a measure of genuine technological leadership. Although even more substantial moves of theis sort are needed, recall that this is what Brzezinski had in mind. In passing, we also must note that U.S. universities train tens of thousands of foreign students and interact in research with hundreds of academic institutions round the world. Such training, while it runs the risk of increasing "brain drain," must be sharpened and sustained.

Finally, connecting all of the major sectors in the United States, the media are important. There are broad deficiencies in the public's literacy about science and technology as well as about international issues. Newspaper and magazine coverage of science has been weak, but it is improving. The general public probably does not understand the future global stakes in maintaining our scientific and technological capabilities. And, of course, foreign technical aid is so extremely unpopular that our relations with the "third world" are fragile and contentious.

INTERNATIONAL PATTERNS: CENTRIPETAL OR CENTRIFUGAL FORCES?

We will touch briefly on international institutions. Here, of course, are many kinds of bilateral connections, as well as multilateral alliances and other coalitions, together with the United Nations.

Bilateral government agreements on science and technology involving the U.S. number many more than even well-informed observers are aware. It took 15 single-spaced pages to list the important ones in a January 1979 survey by the State Department. Beyond these are thousands of relationships involving industrial and academic groups. Comparable statistics apply to most of the other industrial countries and their private institutions. These bilateral partnerships are often the most effective—some would say the *only* effective—ways to get things done.

Yet it is the multi-lateral political-economic coalitions, as well as formal treaties and the military-oriented alliances, that usually receive attention internationally—e.g., NATO, OECD, the Group-of-77, and OPEC. To be sure, there also are many internationally effective nongovernmental associations (such as ICSU), particularly in the sciences, medicine, and engineering. And there are topic-centered groups—say, in health and agriculture—that cross the public/private sectors in rather productive ways, transferring technical skills with high leverage and overcoming tall political barriers with deft pragmatism; e.g., CGIAR.

The United Nations system today is a frustrating blend of idealism, technical commitment, bureaucratic waste motion, and weary rhetoric. Ironically, the reputation of the U.N. seems to be falling as rapidly as the need for its potential action is rising. Gravely disabling political and economic pressures often diminish the U.N.'s once-honored professionalism. Not least in the preparations for the Conference on Science and Technology for Development (held in Vienna during August 1979) we have seen how inter-agency squabbles and vague agendas can erode much of the base for seriously focussed international action.

ILLUSTRATIVE FIVE-YEAR PRIORITIES: WHAT IS S&T IN (FOR) FOREIGN POLICY?

It might be stimulating to assess a few *hypothetical* priorities for a five-year program spanning the roles of science and technology in (and for) foreign policy. Let us take five areas for a brief review.

The *mission-oriented governmental agencies* of the OECD countries possess ample justification for renewed emphasis on R&D. Along with a good case for a clearer definition of the R&D mission of each agency, there also is an increasingly strong argument for explicitly adding international technical activity. For example, since the U.S. and NATO must depend more heavily upon R&D as a hedge in arms control policies, there is great justification for cooperating actively on more R&D—serving goals in arms control, intelligence, communications, and all stages of weapons development.

Some observers argue that the U.S. National Science Foundation should reemphasize its traditional priority on *basic* research in agriculture, reproductive biology, tropical diseases, energy, and other areas; such work would help the U.S. *and* other countries.

In the State Department, as another example, there is cogent justification for more research on many of the policy-problems mentioned earlier.

Overall, most agencies need modest additional funding (even very small sums would go a long way) to support international exchanges, seminars, and short collaborative visits linking scientists from various countries.

We have listed a series of possible increases in support and so we must consider *what might be cut among the existing agencies*. For example, parts of the more conventional "technical assistance" programs of agencies such as AID could be trimmed—*if* there were simultaneously a reorientation toward genuinely collaborative work, building up the capabilities *in* LDCs that can enable them to carry out more self-reliant choice-making about technology. Another area that might be cut is the stage of defense R&D that is intermediate between research and production; very large investments are made in this stage and, of course, only a few systems survive into deployment.

"Global concerns" provide a natural justification for more tightly *linking the activities of the industrialized nations.* Among the current technical topics of unusual urgency would be: climate and water resources; energy and related natural mineral resources; and the broader tasks of achieving a better *joint* analysis of how to proceed on the longest lead-time activities required for modernizing the third world while fostering reasonable stability.

Prospects for greater economic competition in the world's trade can not be blinked away. The growing number of "middle tier countries" will become both potential collaborators and stronger competitors with the industrialized nations. One worthwhile step in this complicated pattern would be to arrange for a new set of investments in *university-industrial linkages within the private sector,* meeting national and international responsibilities that are now drifting toward already overburdened governmental and intergovernmental institutions.

Building meaningful *cooperation by industrial countries with LDCs* will continue to be a major challenge, ultimately determining the degree of peace and progress in the world. There will certainly have to be some accommodations by, and greater respect for, the multinational companies from *both* DCs and LDCs. Taking account of the ways in which science and technology actually are diffused internationally, MNCs are essential. At the same time, there will have to be more sober realism among many developing countries about the likely slow rates of change in achieving their goals.

A point to emphasize in closing is that we have absolutely urgent needs for *new initiatives in policy analysis.* We should have much deeper "cross-national comparisons" among both developed and developing countries, to show the historical roles of technology in modernization. To reach more reliable analysis, we should stimulate a few new centers—preferably on an international basis—that would examine the interactions between science/technology policy and economic policy, within the larger international diplomatic context.

It is of course also true that we need a fresh analysis of the consequences of SALT II with respect to the longer term prospects for meaningful detente, for SALT III, and for actual disarmament. This topic is now more discussed than studied, and it requires a high priority with respect to world-wide trends affecting the independence of many countries.

As a final example of a subject that requires deeper policy-analysis, consider one that might be regarded as too ideological: examining whether governmental, or private, management has been most effective in efficiently steering R&D toward productive social and economic purposes. This subject has obvious implications for both North-South and East-West trends.

For analyzing many policy-issues, we shall have to muster unusual talent and courage. We shall have to explode myths, puncture rhetoric, avoid euphemism, measure what is important to measure, and stand up for the humane principles where technology and foreign policy meet.

Overview of Policy Issues: Panel Report

Panel Chair: HARVEY PICKER
Panel Members: Jean Cantacuzène, R. W. Diehl,
Lionel Johns, Andrew H. Pettifor, Herman Pollack,
Umberto Ratti, A. George Schillinger,
Martin Summerfield, John F. Tormey,
and Frank N. Trager
Rapporteur: Arthur Norberg

SCIENCE AND TECHNOLOGY both influence, and are instruments of, foreign policy. The shared interests and values demonstrated by Russians and Americans at scientific meetings helped to influence the foreign policy of both countries so as to shift from a condition of cold war toward an eventually improved relationship. When deGaulle wanted to demonstrate France's independence from the influence of the United States, he chose scientific cooperation with the Soviet Union as his instrument. It is obvious that the technology of nuclear weapons has had a major influence on foreign policy for many powers, and that an important instrument of foreign policy has been the transfer of technology to Third World countries.

The foreign-policy discussion divided into two distinct segments. In the first segment the panelists added a number of additional issues to the already extensive list provided by Rodney Nichols. The second half of the discussion focused on defense issues in foreign policy.

GENERAL DISCUSSION

One fundamental difficulty requiring analysis is that there is no conceptual framework relating to the nature and role of science and technology in international affaiirs. Up to now international affairs have been the province of the international-affairs specialist, or the economist, who looked upon science and technology as a commodity. It may be that consideration of science and technology separately as commodities may not

---

Harvey Picker is Dean of the Faculty of International Affairs at Columbia University and is Chairman of the Board of the Picker Corporation.

0077-8923/79/0334-0200$01.75/2 © 1979, NYAS

yield a useful analysis or solution of the problem. But if we could focus on how one should think about science and technology in international affairs, individual problems might become more tractable. One approach would be to concede that the emphasis in foreign policy should be the importance of science and technology for policy, instead of the reverse. And, further, that an appreciation of foreign policy objectives can help to determine what role for science and technnology is important. This is especially significant in government circles because the focus is on broad problems rather than on science and technology.

The role of science and technology might be more explicitly identifiable if problems are first classified. One useful scheme is to classify problems into two categories: global and universal. Global problems are those requiring multilateral cooperation for their resolution, perhaps even for their assessment. Universal problems represennt that class for which solutions can be devised independently and transmitted across national borders to be applied in other circumstances.

The nations whose populations live at the poorer end of the economic spectrum regard the free flow of science and technology as a simple matter of equity. They believe that they ought to have access to, and assistance in, the solution of universal problems. But at least two shortcomings exist for implementation of such a classification. First, present applications of the term "equity" imply more than the distribution of goods and probably relate to need, to timing, to the "haves," and to the "have nots." Moreover, confidence in macroeconomic models is declining. Second, even if we assembled a list of problems and identified those nations that deserve the most help soonest, to offer such help would require the other nations patiently to wait their turn. Furthermore, prior agreement would have to be reached on the measurement scale to be employed.

A separate, and possibly similar, framework offered involved regrouping of topics under problem areas. Four problems areas are: energy, productivity, urbanization, and military position. All of these are problem areas at least for the industrialized nations, and for many other nations as well. There is a definite role for science and technology in helping to solve problems in these categories, but the nature of that role was not explored. Many view the well-being of their nations as dependent on their relative advantage compared to others in regard to three of the four problem areas. As a result there is pressure in each of the advanced countries to develop a foreign policy which will inhibit the flow of technology in each of these areas, except urbanization.

Foreign-policy issues and the related roles of science and technology, economics, or any discipline, are changing and will continue to change in response to social changes in the United States. Two prominent examples were cited. One, the contingent of persons espousing a no-growth philosophy is increasing. There is a sharp difference of opinion between them and those seeking further growth. Second, can we expect the United States to be endlessly willing to continue to act as a global policeman?

If these attitudes, favoring no-growth in the United States and the withdrawal of the United States from involvement in political problems outside its borders, become powerful enough to move American foreign policy toward isolationism, there may be a concomitant effect on science and technology within the country. Such a foreign policy may well tend to inhibit some aspects of American scientific and technological development.

Future trends in the world regarding science and technology may also warrant a change in the attitudes of the United States. Differences of view between industrialized countries and less developed countries (LDCs) about the use of previously unusable "free goods," such as space, or about the usefulness of supporting science and technology programs abroad, are already stimulating changes elsewhere. The increased use of space for communications satellites and for other projected purposes, and the consequent crowding of the geosynchronous orbit of space has given rise internationally to a strong difference of opinion. From recent discussions of this matter, it has emerged that international control of space may be imminent, and the United States will then become one of a large number of nations involved in that use. What will be the effects of such restriction?

Another concern is the declining support for American scientists traveling to other industrialized countries and its seemingly adverse effects on science on those countries.

On another topic, a proposition needing analysis, and perhaps agreement, is whether technological know-how—patents and processes—are Everyman's, or are privately ownable. It is probably too simplistic to include all technology in one classfication. The technology of smallpox eradication should be Everyman's. What is less clear is whether the technology of producing large-scale integrated circuits falls in the same category and to what extent national foreign policy should influence the spread of that technology. Probably at the opposite end of the spectrum would be technologies such as the production of nuclear explosives, which are considered neither Everyman's nor privately ownable, but are regarded as important subjects for foreign policy determination.

Broader issue areas such as those affecting trade may also initiate change that has direct implications for the role of science and technology in international affairs. The nature of these areas differs for the industrialized countries and the LDCs. Trade in high-technology products among the industrialized countries is hampered because each country operates with different technical standards. To trade in these products it is necessary to manufacture them using the standards of the country to which they will be exported. Hence, a need exists to develop technical standards for safety, patents, and the environnment, and to consider the effects of these standards on trade.

In the case of the LDCs, the issues are more philosophical and more directly removed from science and technology. One issue in this category was raised in the form of a question: Are American citizens willing to endorse setting aside large amounts of money for research and development in the LDCs? If so, it will be necessary to consider the impacts on American industry and trade as the LDCs become competitors. At the moment, discussion on this topic is complicated by the dichotomy between the approaches of the public and private sectors to the LDCs. In many cases, government policies tend to be far ahead of the private sector in willingness to assist research and development in the LDCs. As a direct instrument of foreign policy, the President has advocated the establishment of a foundation to support technological research and development useful to the LDCs. Opposed to this is the currently growing belief in most American labor unions that technological growth in developing countries takes place at the expense of jobs in the United States. As a result, labor unions have begun to turn their attention to influencing American foreign policy in regard to technology.

A further aspect of the potential need for changed attitudes may derive from a challenge to the way the United States projects world trends. Up to now almost all of our projections have been linear, on the basis that the United States is the leader. Policies are then developed from these trends and from the proposition: How does the United States, as a country that will continue to be the leader, act in the world to maintain its leadership? The Japanese have given us the remarkable illustration that this policy does not always have the anticipated result for the United States. For example, contrary to the Japanese, the United States does not seem to have set up very many of the mechanisms they have so successfully used for more than a century. The Japanese comb the rest of the world for science and technology that is transferable to their own country and learn from the rest of the world, discovering a methodology for adapting it to their own uses. The United States tends not to focus on this practice, but our declining position may well make that a future trend for us.

ISSUES IN DEFENSE AND FOREIGN POLICY

The panel discussion also turned to defense, which is one foreign-policy area where science and technology plays an explicit role. The discussion quickly focused on one aspect of this role: that of development and deployment of strategic weapons systems. The issue here concerned the possibility of informed debate when the scientific information is so complex. The panelists recognized that debate about these systems designs occurs at several levels. At one level, there is often a fundamental debate whether physical laws will allow certain processes to occur, and whether such weapons can exist in a meaningful way on earth or in space. On another level, if the weapon is feasible, the public-policy issues then arise whether it might create a more stable or less stable world if it is implemented by one or more countries; or whether such a weapon should be universally outlawed, as nerve gas has been.

Classification of information about such systems prevents the occurrence of such debates, with a consequent loss of private-sector involvement. There is debate in the scientific community, for example, as to whether there should be classification, with the argument advanced that the interests of the United States and NATO would be better served with less security. It can be argued that the academic and the private-sector scientific communities have a genuine stake in this issue, not only in terms of the scientific research involved in validating science for such developments, but also in educating the citizenry here and abroad. A knowledge of the state of the art and its application, if any, should be a valuable aid in developing strategy for NATO or any other alliance. But to say that the United States should permit a detailed public discussion of such issues requires justification. Such open discussion goes against the grain of any military-security system, and might destroy whatever value the system offers.

Grave responsibilities come with the freedom to promote such discussions. If people are already very uneducated about those areas of science and technology that are of greatest concern to them—namely, those posing very large risks—it will be even harder to educate them about complex and abstruse weapons systems. The difficulty is exacerbated by the wide range of vigorous disagreement among the public as to the views and conclusions voiced by scientists and engineers who are regarded as experts in these fields. The process is further complicated by the need to present the information and foster the debate in a context other than an emotional one. The interaction between science and technology and foreign policy can hardly be more evident than in this field.

One last point suggested that the issue be viewed in a generalized way by considering strategic weapons of the future, which might constitute the potential for increasing or decreasing the stability of the world, in their role of maintaining a continued deterrence-type posture. This involves the question whether a deterrence-type posture, of the type that has kept a nominal world peace for the past 25 years, is a decaying policy position. If it is an obsolescent concept, what are the reasons? Is its obsolescence the result of a real lack of usefulness, or a result of perceiving it as less effective? In either case, more pressure is put on research and development to improve the systems on which the United States depends.

The perception of a less effective deterrence posture, the seeming inability of the United States to select and promote a continuous and consistent foreign policy, the lack of agreement on issues between the public and private sectors, and the recognition of areas of strong disagreement abroad, may suggest a more fundamental difference in American life now as contrasted to 25 years earlier when these phenomena appeared absent. During most of its history the world role of the United States was one of moral leadership, but since World War II, that role has been changed by the influence of science and technology, which has brought about the spread of American products, standards, and material philosophy, and has made the United States a desirable place in which to be educated. But the new material philosophy has lacked the moral commitments perceived in the earlier role. And certain extreme practices have suggested a disappearance from our policies of a moral stance.

The matter has become more complex since the country has not in recent years made use of its attractiveness. It lacks a conscious policy, widely held, for identifying assets that should be put to use for its own national objectives, or to serve whatever another nation's vision is of our national interests and of what we can do for others. This lack of vision is evident in planning for the forthcoming United Nations Conference on Science and Technology for Development, in North-South relationships, and in dealing with partners in the Organization for Economic Cooperation and Development (OECD).

Contrast the present state of affairs with that of 30 years ago. In the 1950s, in foreign policy and the defense establishment, there was a shared body of ideas, concerns, assuumptions, and even competencies in many areas, among lawyers, scientists and economists, about what strategic policy was, what its aims were, and what the United States was trying to accomplish. This shared body of ideas informed and disciplined most of the debates, even among people with different views of foreign policy and strategic defense policy. It may be that the early success of the United States masked or mitigated the disagreements, and as its policies and actions achieved less success, the disagreements moved closer to the center of discussion. Or, more probably, it was easier in the fifties for a people to agree who had recently been united with a common objective during World War II, and who still largely shared the view of a common danger from the cold war.

Certainly today no such unified self-image exists. People now are from different professional backgrounds, with fewer shared experiences under fire in formulating major policy decisions and in confronting the consequences of those decisions. Moreover there is less agreement on the priority of goals and the acceptability of the concommitant costs of achieving these goals.

It would be appealing to be able to conclude from the discussion that a simply stated paradigm of the interaction of science and technology with foreign policy had emerged. The only safe prediction is that the problems and issues of foreign policy will interact even more intimately with science and technology in the future. There are two basic reasons why this is so.

First, as a result of their involvement in weapons development during World War II, their participation in disarmament negotiations, and their direct advisory function in government through organizations such as the President's Science Advisory Council, scientists and engineers have come to regard themselves as no longer being concerned only with problems in their laboratories. Many have come to regard themselves as having a right and a duty to become involved in foreign-policy debate and formulation. The link between laboratory and government in regard to foreign policy has been firmly forged. It will not grow weaker in the future. The link will intensify the pressure to put science and technology at the service of foreign policy, and will heighten the responsiveness of foreign policy to new scientific and technological developments. Increasingly, scientists and engineers will become part of, or move back and forth into, those parts of government concerned with foreign policy formulation and implementation.

Second, as Dr. Nichols' paper indicates, foreign policy is concerned with an increasing number of issues which have a large science and technology component. As a result, science and technology has, in itself, become a foreign policy isssue of prime importance. It is only in the last half of the twentieth century that most countries have come to see their welfare as no longer dependent solely on their factor endowments of land, resources, and population. They have now come to recognize that their scientific and technological capacity can alter their relative and absolute well-being over a wide range of aspects from military power to health and economic status. To safeguard or improve their condition, they believe that they must increasingly direct their foreign policy to science and technology issues. Who will have what access to scientific and technological developments has become one of the most critical issues of foreign policy.

———————

As the interaction between science and technology and foreign policy widens over a larger range of issues, and the importance of these issues intensify, there will be a growing demand for increased knowledge about the interaction. Centers for the study of the subject and discussions such as were held at this conference will not be able to provide pat answers, but they will be increasingly necessary in order that the dimensions of the problem and the costs and benefits of alternative options be, insofar as possible, understood.

Aid to Developing Countries

KLAUS-HEINRICH STANDKE

THE QUESTION OF aid to developing countries presents several fundamental difficulties. In both the domestic context and the international diplomatic world, the word "aid" has particular connotations that closely suggest charity. Besides provoking the justifiable sensitivities of some recipient countries, the term "aid" is misleading. An article in the March 26, 1979 issue of *Time* magazine[1] indicated that the material results of this so-called aid are often significant and mean good business *for the donor country*. For example, according to the same article, for every dollar that the United States contributes to international financial institutions that give aid, the recipients spend two dollars to buy goods and services in the United States. For every dollar paid by the United States into the World Bank alone, $9.50 flows into the nation's economy in the form of procurement contracts, operations expenditures, and interest payments to investors in the bank's bonds. In short, this essay underlines that "the United States does well by doing good."[1-3] Perhaps the term "cooperation" explains better what we have in mind when dealing with the North-South relations.

Another concern is the almost inevitable degree of generalization that accompanies discussions of Third World issues. In any discussion of the use of science and technology for development there is a danger that undue focus will be placed either on the science and technology dimension of this admittedly complex issue or on the aspect of development. There is not sufficient understanding of the needed interaction between the two. Also there is the danger of oversimplification in dichotomizing "developed" and "developing" countries. At the least, a grouping of countries according to their resource endowment, their per capita income, their cultural specificity, their climatic and geographic conditions—to note only some variables—is needed before any sort of conclusive analysis can be made.

A starting point for our discussion should be the generally accepted, athough ill-defined notion that science and technology is an important, if not decisive, factor in the development process of nations. And it is at

Klaus-Heinrich Standke is Director, Office for Science and Technology of the United Nations.

0077-8923/79/0334-0208$01.75/2 © 1979, NYAS

this early stage of deliberation that the difficulties in understanding this complex issue begin.

As Thorkil Kristensen observed:

> Looking at the world today, we find that the most decisive difference between what we call more or less developed countries is that the former are able to apply modern knowledge to a larger extent than the latter. This is why the more developed countries are richer. It is therefore a major aim of the poorer countries to have more and more of this knowledge applied to their various activities in order to reach higher economic levels.

He concluded from this that "it is misleading to talk about 'developed' and 'developing' countries." No country is fully developed, he says, since no country can have all of the possible knowledge applied to all its activities.[4]

The same line of thought is followed by the signatories of the Andean Pact (the so-called Cartagena Agreement). The introduction to their technology policy concept contains the statement: "The stock of available knowledge and its adequate use is a decisive factor in the ability of a society to fulfill its economic and social needs."[5] The international dimension of this issue within the present North-South dialogue is illustrated in the following statement: ". . . the stock of scientific and technological knowledge available to a country, which it incorporates into its production activities, is a power element, and is reflected in the country's relationships with the rest of the world. Technological superiority has political significance and as the key that distinguishes interdependence from dependence, fluctuates continually according to the access to scientific and technical advances, and the degree to which advantages are actually realized."[6-8]

Consequently, while preparing for the forthcoming United Nations Conference on Science and Technology for Development (UNCSTD), member States—particularly those from the Third World—are putting greatest emphasis on the need for all countries to share existing scientific and technological knowledge. This is reflected in the fact that, at the recent third session of the Conference Preparatory Committee, the following item was introduced as Target Area No. 1 of the Conference itself: "Willingness and preparedness to share the knowledge and experience gained by each country through its own efforts to achieve political, economic and social self-reliance are inherent to subscribing to the concept of global interdependence and true cooperation between all countries."[9]

A number of proposals have been made on how to reach this target: (1) creation of international cooperative information systems and networks; (2) facilitation of access of developing countries to the knowledge available in developed and other developing countries; (3) promotion of mechanisms for person-to-person exchange of information; (4) research and development (R&D) by developed countries devoted to specific problems of developing countries; and (5) cooperative R&D.

Although it is relatively easy to make up such a list of demands, which are all meaningful and important, the actualization of such targets does not depend solely on the availability of financial resources and the existence of political will.

Knowledge is not value-free. In the statement from the Andean Pact Technology Policies, an important distinction was made between "production technology" (that is, the knowledge essential to transform inputs into products or services) and "consumption technology" (that is, the characteristics of the goods and services that satisfy or induce requirements at the consumer level). Such technological knowledge tends to reflect the multiplicity of consumer and producer motivations inherent in both the innovating industry and the target clientele; and these are constrained by cultural patterns and historical evolution, by access to capital market size and profit opportunity, by perceptions of what contributes to a good quality of life, and by economic, social or political status.

Consequently, countries importing technologies either for production (that is, how to organize, design, produce) or consumption (that is, what goods will be available to consumers) are simultaneously buying an implicit set of values which may or may not mesh with their own preference systems.

At present, it appears that no single country in the world has been able to solve this inevitable value conflict. In order for developing countries to reach some sort of industrial parity, they must catch up with developed ones. However, industrialization automatically implies the application of scientific and technological knowledge and values. It is the equating of industrialization with development that causes many of the disillusions and misconceptions faced by Third World countries today in their struggle to determine their own economic, social and political fates.[10]

A point related to the issue of sharing knowledge is the question of transferability of "old" knowledge, as distinct from the necessity of generating "new" knowledge. In the assertion of the need for R&D, too much emphasis, according to a recent OECD Workshop on Scientific and Technological Co-operation with Developing Countries, has been put on research leading to the creation of new knowledge, whereas for developing countries, the problem essentially lies in setting conditions that would favor the utilization of knowledge, much of which is already at hand. However, there has been a high entry fee for converting new knowledge into usable technologies, which developing countries have not been able to "pay."[11]

Another interesting idea was advanced at this OECD workshop. Participants appeared to be more of the opinion that the quantum of existing usable knowledge is relatively small and perhaps even declining in the face of new exigencies, including the large unknowns of the physical and social environment in which the poor of the developing world are to be found, and new perceptions of the character and function of technology in a basic human-needs development perspective. If anything, they suggested, the future will demand a far greater capacity to generate new knowledge.[12]

The capacity of developing countries either to absorb or adapt existing scientific and technological knowledge or to create new knowledge better suited to their own needs is the key issue, to which all other issues concerning science and technology for development must ultimately relate.

As E. F. Schumacher stated so well, "Development does not start with goods, it starts with people and their education, organization and discipline. Without these three, all resources remain latent, untapped, potential."[13] He went on to say that

> . . . development cannot be an act of creation, that it cannot be ordered, bought, comprehensively planned; that it requires a process of evolution. Education does not jump, it is a gradual process of evolution. Education does not jump, it is a gradual process of great subtlety. Organization does not jump; it must gradually evolve to fit changing circumstances. And much the same goes for discipline. All three must evolve step by step, and the foremost task of development policy must be to speed this evolution. All three must become the property not merely of a tiny minority but of the whole society.[13]

To quantify the order of magnitude of this problem, let me quote a few figures. Of the more than 4 billion people on earth, some estimated 800 million adults are illiterate. Practically all of them live in developing countries. Furthermore some estimated 250 million children of school-age are not enrolled in schools and practically all of these children live in developing countries.[14]

Of more immediate relevance to the absorptive or generating capacity of developing countries is the distribution of R&D expenditures and manpower. The latest available figures date from the year 1973. Of the estimated total world expenditures for research and development amounting to 109 billion dollars, only 2.3 percent has been spent in developing countries. The ratio in scientific manpower is slightly more favorable: of the estimated total world figure of 4,950,000 scientists and engineers, some 12.0 percent live in developing countries.[15]

To balance this unsatisfactory situation, a number of political, philosophical or simply "pragmatic" approaches have been put forward by the developed countries, by the developing countries themselves, and by both groups. Most prominently, the developing countries proposed the concept of the "New International Economic Order," which was accepted during the VIth and VIIth special sessions of the United Nations General Assembly by all member States. In addition, there is the "World Plan of Action for the Application of Science and Technology to Development,"[16] which the Advisory Committee on the Application of Science and Technology to Development prepared for the Second United Nations Development Decade. Furthermore there is the series of United Nations Conferences and of conferences organized by the specialized agencies of the United Nations system. Each of them resulted in recommendations, plans of action and resolutions that have particular salience to science and technology and their relationships to development.[17] A vision of self-reliance or of collective self-reliance is held by developing countries.[18-21] This concept has been introduced by some Third World leaders like President Nyerere of Tanzania. It had been articulated earlier, in political terms, by Presidents Nasser, Nehru, and Tito, who founded the movement of the "nonaligned" states. The rather idealistic idea of self-reliance has been put, by means of the multilateral negotiation machinery of the United Nations, into a program framework of economic cooperation among developing countries and of technical cooperation among developing countries.

The most comprehensive political framework of all these concepts is the one called the New International Economic Order (NIEO). All North-South discussions at present on the world agenda should be seen under this rubric.

The NIEO has a number of aspects dealing with science and technology:

(1) the cooperation of developed and developing countries in the establishment, strengthening, and development of the scientific and technological infrastructure of developing countries;

(2) the significant expansion of the assistance of developed countries in direct support of the scientific and technological programs of the developing countries, in accordance with feasible targets to be agreed upon;

(3) the substantial increase of the proportion of the research and development in developed countries devoted to the specific problems of primary interest to developing countries and to the creation of suitable indigenous technology, also in accordance with feasible targets to be agreed upon;

(4) the expansion of international cooperation on the basis of principles and regulations designed to adjust the scientific and technological relationships among States in a manner compatible with the special requirements and interests of developing countries, especially in the transfer of technology.[22]

In contrast to the developing countries' proposals for an NIEO, the developed countries did not introduce a comprehensive view, concept, plan or program that would define a world development guaranteeing a sustained growth for all countries of the world. This is, I believe, a severe omission and it makes the ongoing North-South dialogue even more difficult.

Of course, there are numerous, rather academic attempts by the Western industrialized countries to look into both global issues (which includes the development of developing countries) and partial aspects of development through the application of science and technology with particular emphasis on developing countries. Examples of these attempts include the World Problematique of the Club of Rome, [23,24] the RIO-Project of Professor Tinbergen,[25] and the Leontief model of world development.[26] Further, concepts have been advanced to re-emphasize the quality of life,[27] to meet basic human needs, particularly in developing countries,[28-32] or temporarily to delink certain developing countries from the general trends of development in the world.[33,34]

We have to accept that the leaders of developing countries regard with suspicion such concepts aimed at "helping" developing countries. We have to accept that the development, for multilateral negotiation within the United Nations, of appointed spokesmen for groups of countries (the Group of 77 for developing countries, Group B for the Western industrialized countries, and Group D for the eastern socialist countries) may have favored group bargaining positions that are largely based on mutual mistrust.

What is now needed is an emotion-free, objective, disinterested assessment by all concerned—North, South, East, and West—of what we mean by "development." Only when this is better understood will we be able to consider in depth which mechanisms for aid and cooperation can be mobilized to achieve the desired end. Only one of these mobilizing factors will be science and technology.

I am pleased to see at this particular time that because of the worldwide preparations for the United Nations Conference on Science and Technology for Development, the topic of our working group at the present Conference has received widespread attention. What worries me, however, is that the level of expectation of what science and technology can do for national, regional and world development has been raised to a point almost beyond reach. Some observers feel that the disillusions after Vienna may be damaging to the science and technology community in many countries—developed and developing—for a number of years, unless it can be demonstrated not only *what* science and technology can do, but also and more importantly, *how* this can be done. To achieve this, discussion on the topic of science and technology for development should not take place only within diplomatic United Nations gatherings.

It is also necessary to reach out by revising existing bilateral governmental relations so that a visible interest in science and technology will be maintained. In addition it is important to reach out to industry and to academia. It is difficult, if not impossible, to foster a new understanding of the role of science and technology for development if the nongovernmental community, which is vital for any "action implementation," is not fully aware of a developing country's sensitivity to a paternalistic development concept proposed, even with best intentions, by a developed country.

Instead of offering easy conclusions, I would like to present—in the form of questions—some of the underlying psychological obstacles that make the current debate on science and technology for development so difficult:

1. What developing countries do we mean when talking about the Third World? Do we sufficiently differentiate the quite different characteristics of these countries?

2. Which "development" do we have in mind when offering "aid" through science and development?

3. How long does it take to create an adequate science and technology infrastructure in developing countries so that they are enabled to act as equal partners in the present North-South dialogue and to define their own interests? What can we do to accelerate this process.

4. What do we mean by "appropriate technology" for the developing countries?

5. What are the arguments in the controversial "basic human needs" debate?

6. Why is the "technology transfer" issue of such emotional importance?

7. Do we mean self-interest when we say "aid"?

8. Do we believe in an egalitarian concept of "world development," in which we want to share our skills and our knowledge? Or are we pleased with the "status quo," which we seek to preserve by "buying time" through some form of charity?

9. Should a common concept of science and technology for development be held by all developed market-economy countries (for example, through the OECD)? Or should each government continue to tackle this issue in its own way?

10. Are we losing sight of the pressing issues facing all mankind because the present world debate is focusing almost exclusively on North-South issues?

It is hoped that a better understanding of these issues will lead to more productive and constructive policies for the application of science and technology for development.

REFERENCES

1. TIME. The downs and ups of foreign aid. March 26, 1979: 52.
2. SOMMER, J.G. 1977. Beyond Charity. Washington, D.C.
3. MAZOUI, A.A. 1979. The cultural aspects of foreign aid—Reasons for aid: Charity, ideology or self-interest? In Development and Cooperation 1: 6–11.
4. KRISTENSEN, T. Development in Rich and Poor Countries. :13. New York, N.Y.
5. Andean Pact Technology Policies. 1976. Junta del Acuerdo de Cartagena. Ottawa, Canada.
6. Ibid. :5.
7. KELLER, O. & F. DÜRRENMATT. 1974. On the interaction between science and technology and political power. In Die Physiker. :44, Munich.
8. STANDKE, K.-H. 1970 Europäische Forschungspolitik im Wettbewerb. :39. Baden-Baden.
9. UNITED STATES CONFERENCE ON SCIENCE AND TECHNOLOGY FOR DEVELOPMENT (UNCSTD). Preliminary Draft Plan of Action A/CONF. 81/ PC. :7.
10. STANDKE, K.-H. 1977. Utilization of new technologies in developing countries. Materials and Society 1: 46.
11. OECD Workshop on Scientific and Technological Cooperation with Developing Countries, April 10–13, 1978, SPT (78) 17. :5.
12. Ibid. :5.
13. SCHUMACHER, E.F. 1975. Small is Beautiful. :168–69. New York, N.Y.
14. McHALE, J. & M. McHale. 1977. Basic Human Needs. :33. New Brunswick.
15. GERMIDIS, D., N. JEQUIER & M. Brown. Technology and the Development Process. :5. OECD, STCD/S/78.4.
16. United Nations World Plan of Action for the Application of Science and Technology to Development. 1971. New York, N.Y.
17. LOGSDON, J.M. & M.M. ALLEN. Science and Technology in United Nations Conferences. A Report for the U.N. Office for Science and Technology, Washington, January 1978.

18. CHAGULA, W.K., B.T. FELD & A. PARTHASARATHI, Eds. 1977. Pugwash on Self-Reliance. New Delhi, India.
19. STANDKE, K.-H. 1977. Science and technology innovations: Self-reliance and cooperation. *In* Proceedings of the Eighth International Conference of the Institute for International Cooperation. Ottawa, Canada.
20. GALTUNG, J. 1978. Development, Environment and Technology towards a Technology for Self-Reliance. Geneva, Switzerland. UNCTAD.
21. RAGHAVAN, C. Towards a new international economic order through collective self-reliance and a strategy of negotiation and confrontation. In IFDA dossier No. 5, March 1979.
22. UNCSTD. Preliminary Draft Programme of Action, *op.cit.* :3.
23. MEADOWS,, D.L. 1972. Limits to Growth. New York, N.Y.
24. MESAROVIC, M. & E. PESTEL. 1974. Mankind at the Turning Point. New York, N.Y.
25. TINBERGEN, J. (Coordinator). 1976. Reshaping the International Order. New York, N.Y.
26. LEONTIEF, W.W., et al. 1977. The Future of the World Economy: A United Nations Study. New York, N.Y.
27. CLEVELAND, H.. & T.W. WILSON, JR. 1978. Human Growth: An Essay on Growth, Values and the Quality of Life. Aspen, Colorado.
28. McHALE, J. & M. McHALE, *op.cit.* :33.
29. INTERNATIONAL LABOUR OFFICE. 1976. Employment, Growth and Basic Needs. Geneva, Switzerland.
30. INTERNATIONAL LABOUR OFFICE. 1977. Meeting Basic Needs. Geneva, Switzerland.
31. INTERNATIONAL LABOUR OFFICE. 1978. Technology Employment and Basic Needs. ILO Overview Paper prepared for UNCSTD. Geneva, Switzerland.
32. SINGER, H.H. 1977. Technologies for Basic Needs. International Labour Office. Geneva Switzerland.
33. SENGHAAS, D. 1978. Alternative Entwicklungstrategien: Dissoziation und autozentrierte Entwicklung. *In* Internationale Entwicklung, No. 3:27–46.
34. MOREHOUSE, W. 1979. Technological Autonomy and Delinking in the International System: An Alternative Economic and Political Strategy for National Development. UNITAR. New York, N.Y.

Overview of Policy Issues:
Panel Report

Panel Chair: NORMAN TERRELL
Panel Members: J. C. Agarwal, Dominique Akl,
Richard Braunlich, J. E. Goldman, Bernhard Hartmann,
Lewis H. Sarett, Ralph H. Smuckler, Charles Weiss,
Jean Wilkowski
Rapporteur: Robert Stibolt

It is generally agreed that the term "aid," as in "foreign aid," is often misleading since those nations offering the aid usually profit by expanding their export markets and by gaining additional sources of raw resources. However, the participants of the panel believe that there is nothing fundamentally wrong with this term since economics is not a zero-sum game and to believe in altruism free of any underlying motives is politically naive.

The real problem confronting the lesser developed nations is seen as one of insuring that their broad developmental interests will be taken into account in any international economic cooperation. For example, how can "production technology" be dissociated "consumption technology" is the technology transfer process?

The question arose as to whether there is an organized effort to help countries to define their own objectives (through, for example, the United Nations). (Just asking this question identifies the glaring weakness of the nations in the "Group of 77"—their inability to analyze their own problems without potentially biased outside help.)* Without this capability, true *quid pro quo* economic cooperation becomes most difficult, and out of this continuing difficulty arises the mutual mistrust characterizing present "north-south" relations.

Considerable discussion focused on the forthcoming United Nations

---

* The Group of 77 is a caucus of developing countries that are UN members. While 77 countries were involved in the original group in the mid-1960s, over 100 countries are now members.

Norman Terrell is Deputy Assistant Secretary for Science and Technology, Department of State.

0077-8923/79/0334-0217$01.75/2 © 1979, NYAS

Conference on Science and Technology for Development (UNCSTD) held in Vienna, Austria, August 20–31, 1979. The outcome of this conference cannot be anticipated, for such conferences are often complex, confused, and highly political. It is thought that the forthcoming conference is likely to be too highly politicized for constructive discussion, and that more attention will be paid to ideological than to pragmatic issues. However, the conference will probably serve to force both the United States and all other conncerned parties to define more clearly their policy towards science and technology for development.

The relationship between technical capabilities and balanced growth within the developing nations is another key area for consideration. A major conclusion of the discussion was that without a social and economic infrastructure to ensure education and technical training of the indigenous population, technology transfer and integration cannot occur. The panelists generally thought that the leaders in many of the world's developing nations have not *fully* appreciated this critical requirement. The inevitable result has been that nationals often manage foreign technologies within the borders of their own developing countries primarily for the benefit of the foreign nation, rather than for that of the recipient country.

Approaches to the techniques and mechanisms of technology transfer raise many controversial questions. Is it enough to transfer "old" technology to the less developed countries, or should "new," more culturally specific or appropriate technology be developed for them? The first response was that application of "old" techniques in a new environment often requires "new" knowledge in order to effect adaptation to new circumstances. However, unless there is a care of technically trained people within the less developed country, there is no mechanism to accomplish this. One panelist expressed the view that people in less developed countries must develop the capability to evaluate the prerequisites for utilizing technology within their borders if they are to solve the "brain drain" problem. Another aspect of the "old versus new" question is that some problems required entirely new knowledge, such as that called upon to develop drugs to combat tropical diseases not found in industrialized nations. Cooperation on this front is as profound a problem as that of transferring "old" knowledge.

The broad mechanisms for technology transfer are also receiving increased attention. Some Fortune 500 companies have larger assets than do many developing nations. This raises the possibility that industry-to-government rather than government-to-government links may be a better mechanism for technology transfer, since comparably sized organizations can interact on a basis closer to equality. It was pointed out that despite the heavy criticism multinational corporations have received for their involvement in international politics, their record with respect to technology transfer is far better. It was widely agreed that the mind-set of many foreign bureaucrats and diplomats in less developed countries must be changed so that they can recognize the benefit their countries have derived from multinational corporate activities. Here, as elsewhere, psychology plays an important part. New ways should be explored to inform national planners of the role multinational corporations could assume in technology transfer.

The transfer and use of technology are highly complex processes that are affected by numerous restrictions, qualifications, and opportunities. Some of the principal dimensions of this issue were touched upon in the panel session. How this process, from transfer to acquisition to integration, can operate more effectively for all parties at interest will remain the underlying momentum influencing the development of new approaches and policies to meet the needs of the coming decades.

Opportunities for Cooperation Between Government, Industry and the University

LEWIS M. BRANSCOMB

"REINVENTING" CARS GAINS MORE BACKING

Transportation Secretary Brock Adams was nearly laughed out of Detroit three months ago when he called on the nation's automobile industry to "reinvent" the car. He was not quite sure what he was appealing for, he acknowledged, and industry executives were even less certain.

This week, two Congressional committees staged hearings on some variation on the theme of reinventing the car. And today, several key automobile industry executives praised the suggestion as one of the best things to come along since sliced bread.

Several members of Congress have supported Secretary Adams' suggestion that the Government put hundreds of millions of dollars into new basic research toward "reinventing" the automobile through joint government-industry projects.

A hearings today called by the Senate Committee on Commerce, Science and Transportation, one Representative called for the establishment of a federally chartered research institute and the appointment of a White House assistant for automotive affairs, while another proposed raising the ante higher by creating an Apollo-type of project. Hearings were held earlier this week by the House Subcommittee on Transportation, Aviation and Communications.

Mr. Adams, meanwhile, has again postponed his much talked about automotive summit, at which a working plan could be developed for a joint government-industry effort.

He proposed such a gathering of automobile executives and himself in December during his visit to Detroit to speak before the Economic Club of Detroit. In that talk, and a much tougher text, he roundly criticized the industry for failing, in his opinion, to make any meaningful improvement in the automobile in the last 60 years.
— Reginald Stuart in *The New York Times* (March 25, 1979)

THIS ARTICLE presents in microcosm the issues in the evolving relationship between government and industry with respect to

---

Lewis M. Branscomb is Vice-President and Chief Scientist of the International Business Machines Corporation and former Director of the National Bureau of Standards.

0077-8923/79/0334-0221$01.75/2 © 1979, NYAS

specific technologies. I call your attention to the following features of the story:

(1) A government official (Secretary Brock Adams) castigates the industry for "not making any meaningful improvement in the automobile in the last 60 years," Thus performing a feat of hyperbole clearly intended to occupy the political high ground without committing him to anything. Nevertheless, his suggestion that it is time for government to take the lead in "reinventing the car" is a call for increased cooperation between government and the auto industry.

(2) Industry officials, initially hostile to any government role, appear to shift toward support of government sharing in "the burden of this expensive advanced research." The reporter suggests increasing industry acceptance of a major government role in the specification, design, and development of radically different kinds of vehicles and power plants.

(3) But, notice that every business leader who was quoted making a favorable comment on Adams' proposals coupled that comment with an appeal for relaxation of federal laws and regulations. According to the report, Chrysler's S. L. Terry specifically proposed that the government "should establish an independent authority to assemble and assess all regulations that deal with the automobile." W. W. Slick, Jr., of American Motors, urged relaxation of antitrust constraints that preclude industry cooperation on basic research efforts.

(4) Notwithstanding that congressmen called for "Apollo-type projects," special federal research institutes, and a special assistant to the President for Automotive Affairs, the auto industry was actually welcoming only basic research. Philip Caldwell of Ford was quoted as saying the government's role was in basic research, "for example the combination process, fuel cells, materials and alternate fuels." That is a long way from "reinventing the automobile."

The journalist leaves us with a rather comfortable feeling. The hostility that usually seems to characterize the relationship between government and industries that are subjected to strenuous regulation is portrayed as dissipating. Some readers might even get the impression that an agreement to bury the hatchet and let the government take the lead in "reinventing the automobile" was just around the corner. I submit however, that the Congress and the industry—as described in this story—were talking past each other.

Political leaders are increasingly under pressure not to leave product design and technology solutions to the market-driven judgment of the private sector in situations like this. No one contests the appropriateness of the Federal role for the support of basic research and for the specification of performance standards for fuel economy, pollution levels and safety. But a direct role by government in choosing and financing commercial technology is much more controversial, since it is generally recognized that cost discipline and consumer acceptance are best attained when technology is developed in a commercial environment. As a compromise, government often turns to demonstration projects. But the concept of the government demonstration project is often based on the idea that the best way to encourage technological change is a visible demonstration of working models of alternative technologies. This is often not the case.

This preference for the massive Federal demonstration expenditure in areas of evolving technology is visible in many areas: the solar lobby's desire for demonstration plants, the Personal Rapid Transit demonstration in West Virginia, and the public data satellite demonstration of information services for remote areas.

Last year, John Young of the Congressional Office of Technology Assessment prepared a study under the guidance of an R&D panel I chaired. This study examined the demonstration program as a tool of Federal policy when the government wishes to accelerate development and acceptance by the public of new technology that must ultimately be delivered to the public by the private sector. I provided a set of criteria to guide legislative committees on R&D strategies to be used to achieve public ends through private application of R&D. I commend the study to your attention.

Because the examples of demonstration hardware are more familiar and easier to deal with, the study emphasized "soft," or social, demonstrations. The heart of the recommendation is, of course, that if a demonstration is to be undertaken, the government should identify in advance what it is that one wishes to learn that cannot be learned more cheaply and quickly some other way, and then build into the demonstration the measurements and evaluation that insure that some objective, other than simple political visibility, will be attained.

My second observation is that the words "basic research" fail to communicate the intent of the industry in this discussion. I am sure that the automotive leaders are not opposed to the government funding what the National Science Foundation (NSF) means by "basic research." I am equally sure that that is not what the industry leaders had in mind. Remember that Caldwell referred to fuel cells, alternate fuels, and the like. That is not the kind of work that gets done in a useful way without being guided by and in good contact with the works of product and process design and engineering. Indeed, as the industry leaders called for "basic research" they also called for a voice in the research strategy. I quote again from the *The New York Times:* "They also made it clear that they felt themselves well suited to having a major, if not controlling, hand in the government's proposed efforts . . ." and Mr. Caldwell is further quoted as having said, somewhat plaintively I imagine, "It is important to keep research goals in perspective. It is not enough to solve the technological piece of the puzzle. We've also got to satisfy the customer."

Technical people, perhaps more than anyone, are beguiled by suggestions like that of Congressman George Brown that a Federally chartered automotive research center ". . . would enable us to pool all the public and private resources of this nation to develop energy-efficient cars that run on some other fuel than oil." I assume they would be nonpolluting too. The question is, how can this result be achieved, antitrust laws notwithstanding?

Joint government-industry development of consumer products is a new area for the United States. When the government undertakes a technology demonstration, it can certainly succeed in demonstrating that new kinds of vehicles can be built. Most demonstrations undertaken by government are very conservative technically and venture little, but I can imagine that in this field a set of daring projects might be attempted. The question is: How can such vehicles, together with the entire system of fuel generation, distribution, vehicle maintenance, traffic control, and so forth, be made financially accceptable to the public and how can the transition of infrastructure from the current system to the new one be managed? How can a government agency expect to achieve this if it lacks the experience, talent and resources of the industry, yet insists on calling the tune from Washington?

A large part of the technology investment in any manufacturing enterprise, surely in automobiles, goes into engineering the product and the production tooling and processes to make the vehicle manufacturable. People may not regard today's autos as bargains, given recent price rises, but they have little concept of the fine-tuning of the automated production facilities that permit the cost to be no higher than it is. For a "reinvented automobile" this can be a vast investment. In any case, the production and materials engineering must be done hand in hand with the product design. No Federal demonstration is likely to be capable of dealing effectively with this interface. Indeed, the best evidence I know of, to prove the inadequacy of technology demonstration, is the Soviet experience with the Ryad series of computers. In 1972 the equivalent of a government R&D laboratory built six fully working production prototypes of the EC 1050 computer. To this date they appear not to have been successful in manufacturing this quite workable design in factories geographically and organizationally remote from the laboratory.

At the other extreme, however, the basic research in physics and chemistry supported in our universities by the NSF, while very helpful to research scientists in the automotive research centers, is not all the government can or should do to help shift automotive technology to safer, less polluting, and more energy-efficient technology. In fact, two kinds of programs are useful and appropriate in an example like this. The first I will call federal support of exploratory, generic technnology (this is what Dr. Delapalme called "fundamental technology"). This is just the kind of work Philip Caldwell identified. The second is cooperative development of product prototypes, undertaken jointly by government and industry using the concepts of the Cooperative Agreements Act of 1978. Let me discuss these two concepts in a little more detail.

Exploratory Generic Technology

This work lies in the hazy domain between what the science departments of universities regard as most worthwhile, on the one hand, and the product or prototype development project aimed at specific operational performance criteria on the other. I don't dare call it "applied research," because those words are now devoid of any residue of meaning. But it is the kind of work that scientists, rather than engineers (except those with PhDs), are most likely to do. It is the kind of work that best fulfills its purpose when it is published rather than held proprietarily. But it is done for a well-understood reason. There is an identified need, and the researcher gets satisfaction from seeing that need fulfilled. This is the kind of work that finds its way into productive use on a time scale of about 2 to 7 years in most industries. It is what the bulk of corporate central research laboratories are all about—exploring phenomena, materials, and new systems concepts in order to expand and illuminate the technological choice open to the engineers in the profit-center parts of the enterprise.

How can it be that this most useful, yet at the same time most intellectually satisfying kind of work—economically valuable yet fully capable of staving off the threat of perishing that threatens academics—how can it be that work whose value has a time constant of about one presidential term gets so little coherent support and attention in our government? The answer lies in an examination of the constituencies for it, and their behavior toward Federal help.

The academic constituency for exploratory general technology has not been a strong voice, partly because it is drowned out by those advocating the support of "pure" research, and partly because industry has abdicated its responsibility to share its visions and dreams for the technical future with colleagues in academia. A combination of NIH ("not invented here") and too much opulence on both sides in the 1960s delayed the now-evolving new relationship between universities and industry.

But the key element is yet to be solved. How are the strategy and the priorities for this exploratory generic technology to be set? George Pimentel (deputy director of the National Science Foundation) directed just this question at N. Bruce Hannay, Vice-President, Research and Patents, Bell Laboratories) at a recent American Physical Society meeting in Chicago. Hannay and I have been advocating this general notion for quite a while, and he acknowledged that the experts in industry must indeed be given a major voice. But he also cautioned that the job of picking the areas of work must not be left entirely to industry, for industry people can be victims of tunnel vision, too.

I submit that it is time for the mission agencies and for the NSF to try to make this process work. The key is to be very careful how the industry experts are picked. One must remember that the scientists in the corporate research lab—especially those most famous in university circles—are often quite inexperienced and poorly informed about the priority needs of future technologies. On the other hand, engineering managers who work close to the constraints of the marketplace have trained themselves by hard experience not to venture too far beyond well-traveled technological territory. But I do believe that given the right team of people drawn from industry, government labs, and universities, a set of goals specified in technical terms can be established to guide Federally funded work that itself may be conducted in any of these institutions or in combinations of them. It is less clear whether such a program of work can be established in the NSF. Perhaps a broader reorganization of the executive branch science and technology agencies is required to sustain it.

COOPERATIVE DEVELOPMENT OF PRODUCT PROTOTYPES

The second type of activity is much more speculative and difficult to manage properly. The idea is that a government laboratory and a company (several are possible if the Department of Justice would permit) team up to develop advanced requirements set by societal needs but simultaneously to be successful commercially by meeting the needs of individual future purchasers. We have examples where this has worked, usually when the government was going to be an end-user and knew how to determine the design criteria or performance specifications.

Civil aviation offers the best examples since military versions of planes often served as civilian prototypes in early days. When the government was not to be the primary customer, this has not worked so well, although the nuclear power industry's present crisis cannot be laid entirely at the door of the Atomic Energy Commmission.

But the Cooperative Agreements Act of 1978 offers an approach that is worth examining very seriously. For the concept is of an arrangement between government and a non-Federal entity, let us say a company, in which the objectives are set jointly and can be modifiied with experience by mutual agreement, and in which each side takes risk and invests significant resources, both human and financial. Here one can conceive of the government's having a major role in setting the pollution and fuel economy and safety performance goals, and bringing to the partnership the scientific resources of a major national laboratory. The company will bring experience in design, production, and test technology and the awareness of both customer attitudes and the system infrastructure that together constrain the rapidity of revolutionary change. One might hope that in this way reliable estimates of costs might be associated with high-risk technical and market explorations.

Is this an empty dream? Perhaps it is. It is not clear that large institutions in our society can really take public risks like this with impunity, even though risk is shared. It is not clear how competitive fairness is to be sustained; government cannot work with everyone. It is not clear that such investments of public funds would not displace private funds to some extent. Finally, it is not clear that Congress will allow a fairly chosen company partner of government to benefit commercially from success in such a project. And if that is not permitted, the whole purpose of the endeavor is frustrated, quite apart from the fact that companies will not participate on that basis. Nevertheless it may be worth trying.

I have used the automobile as my example, but energy is clearly the highest priority test case. Synthetic fuel production has been the first major test of the cooperative agreements concept, and the President's proposed handling of domestic oil decontrol will put to the test the Department of Energy's ability to expand coal oil and gas technology on a massive scale. If the governmment is unwilling to reinvest in new technology, exploration and development, then the Department of Energy must develop more effective ways of redeploying the tax monies that are "windfall" to the government in cooperative ventures with the industry that has the skill to address the issues.

Roles for Private-Sector Initiatives

A. E. PANNENBORG

IN ANY CONSIDERATION of private industry as distinct from other sectors of society, it is important to recognize as an underlying premise that "private industry" is a rather heterogeneous entity. In the first place, it is certain that, with regard to science and technology policy within industry, significant differences exist between different fields of industry. Secondly, there are systematic distinctions to be made between large and small enterprises in their policies and behavior. And finally, the context within which private industry operates varies for different geographical regions, such as North America, Japan, Europe and the less developed countries. Though the notion of "industry" is usually applied exclusively to manufacturing industry, service industries comprise an increasingly important part of large conglomerate enterprises, and should also be included in the category of "private industry."

OBSERVATIONS ON THE CURRENT SITUATION

A danger underlies the assumption that more research and development (R&D) is always the proper direction to promote. Quite apart from the economic restrictions applying to private industry, progress in science and technology over the course of time generally comes in S-curves. Accordingly, there will always be certain fields in which applied science and technology are entering a phase of diminishing returns. Two examples illustrating these saturation effects are plastics and magnetic bulk materials. It is desirable that management, and in particular R&D management, recognize on a timely basis the onset of a phase of diminishing returns in a certain area and react accordingly so as to shift appropriate resources. The problem is a more difficult one for enterprises active in only a single field where the possible conclusion might be to reduce the R&D potential. Then the pressing question of diversification and its desirability presents itself. Often this might lead to diversification through acquisition, a topic we will return to later.

One can observe that in general the cost of innovation is on the increase in the manufacturing industry. This is not only an instance of

A. E. Pannenborg is Vice-Chairman, Board of Management, of the N. V. Philips Company, Holland, and Chairman of the Board of the Foundation "Toekomstbeeld der Techniek."

0077-8923/79/0334-0228$01.75/2 © 1979, NYAS

S-curve behavior, which can be described as an intrinsic effect, but is also brought about by the external world through regulation. The most extreme case is presented by the pharmaceutical industry where, because of vastly increased time-spans required for registration, the rate of innovation is bound to decrease. It is interesting to note that if we go one step further than regulation and look at planned economies, the extent of basic invention in these countries, with Soviet Russia as the conspicuous example, is virtually nil. One may conclude that invention thrives best in a climate of freedom.

Recently, much attention has been given to the long waves in the world economy, the so-called Kondratief cycle. If this phenomenon is the effect of certain causes, it seems fairly certain that progress in science and technology alternating between various major fields and disciplines is one of the underlying causes.

Science and technology are based predominantly in the highly developed continents. If we compare the three main areas—North America, Japan and Europe—it is striking to find that very few basic inventions have come from Japan. The Japanese seem to be outstanding in follow-up innovations but much less so in breaking truly new ground. The head of the strategic planning of Shell in London, P. Wack, has recently pointed out the tendency in Japanese culture for competence to be valued more highly than originality.

If I may turn to the jargon developed some ten years ago of "technology push" and "society pull" it seems that generally the role of technology push is retracting in the fields of manufacturing processes, components and semifabricated products. For end-user products both for the consumer and the professional, demand pull plays an increasingly important role. If we look at the spectrum of basic disciplines from which technology push is derived, the dark horse in the race is still biochemistry.

If in light of the last praragraph we look at the behavior of the three continents in the field of science and technology, we might say that Japan relies mainly on a hit-and-miss mechanism, that in Europe the emphasis still is primarily on understanding basic data and phenomena and moving only after insight has been gained, and that North America takes a middle position, with the additional mechanism of a far higher rate of creation of new enterprises than in the other two continents.

MECHANISMS OF PRIVATE INDUSTRY

A private enterprise acts on egocentric grounds: it wants to grow as an organization or at least maintain its position by serving society in a way profitable to itself and to society. Any policy of an enterprise, including its science and technology policy, is subservient to that purpose. Private enterprise, as presented in this description, which I think is realistic but does not earn much sympathy in certain circles, is made more acceptable by the custom in many countries whereby experienced senior executives from private industry serve the community in opinion-forming committees of all kinds. If in this context we try to formulate some general observations on the shape of science and technology policy of the individual firm, we can make the statement that the problem frequently presents itself as the choice of proper balance between long-term investment and short-term benefits.

The New Shortages

What we have just said leads us immediately to the controversial role of private enterprise in an era of new shortages: shortage of energy, shortage of food, shortage of raw materials, shortage of unspoiled nature, and shortage of clean air and water. Energy provides a good example. It is generally accepted that energy alternatives have to be developed. On the other hand, almost any alternative proposed is still a very long way from being competitive with oil at its present price. The discrepancy lies for various alternatives between a factor of four and a factor of one-hundred. The situation is aggravated by the observable fact that the general public does not think in terms of total cost of ownership, that is, it generally is not prepared to make a higher initial investment which can be traded off against lower running costs later. This situation then leads to the demand for and recognized necessity of government funding for present-day R&D in this field.

Large Versus Small Enterprise

It seems proper to insert a special word on large enterprise as compared to small or newly founded enterprise. If we make the distinction between discovery (scientific) and invention (technical), we observe that the majority of discovery is done in academia. With regard to inventions, several studies confirm that of the more basic ones not more than 50 percent stem from large industry, though the fraction of total R&D capacity under the wings of the large companies is significantly larger than 50 percent. It will be immediately evident that this 50:50 distribution is not uniform for all fields. It depends to a large extent on the nature of the subject area and the complexity of equipment and the size of teams needed. Philosophically, I think it is highly satisfying that the private inventor still has his place under the sun and I expect and hope that this will remain so. One of the limitations of the large enterprise is its unavoidable focus on established interests with a consequent defensive attitude in its R&D departments.

Living in an ever-changing world, we can make another observation of a certain limitation of the large organization in science and technology. I think all will agree that large R&D departments are virtually immobile. Their strength lies in a well-oiled combination of experienced talent. Any move which would cost the loss of an appreciable part would imply a loss of institutional identity. This further implies that older industrial R&D organizations do not always operate geographically in the optimal environment as small and new industry can, as illustrated by the geographical dislocation of the integrated-circuits industry with its center of gravity in "silicon valley" on the west cost of the United States and the three leading big brothers, Bell Laboratories, Texas Instruments, and IBM largely elsewhere.

Areas of National Concern

We now come to issues where the interaction of science and technology with areas of broad national concern is receiving prominent attention.

The Economy

The short-term time constant of the economy is incommensurate with the time constant which goes with industrial R&D organizations. The hiring and firing of scientists and engineers in step with the economic situation kills the reputation of an enterprise as an employer. This is another case in which the difficult problem of the balance between long-term interest and short-term benefits presents itself. For the large enterprises, one of the most difficult tasks is to anticipate at the right moment the new trends in science and technology that are worth pursuing. My predecessors at Philips did well in this respect by engaging in solid-state research as early as the second half of the thirties.

Sometimes the theory is put forward that the material investment, an enterprise should behave anticyclically with its R&D capacity. I think that while the theory is not incorrect, it is not practicable. The recipe is to try and be as continuous in capacity as circumstances allow.

A point that cannot be omitted here because it has received attention under economic stress is government funding for R&D in private indus-

try. In some fields and many countries this mechanism has achieved a
size which causes very significant distortions of international competition.

Regulatory Policy

With the flood of regulations that has recently engulfed us, I find myself
torn between my roles as coorporate executive and private citizen. Undoubtedly, many regulations are somewhat clumsy and present new obstacles to the success of private industry. On the other hand, with the
growth of populations and the enormous increase in material wealth in
many areas, the world has truly become too small to stick to the axiom:
Free for all. Furthermore, people of this generation have always been
obliged to live with a sizable package of regulations. It seems that man
accepts the regulations which already exist when he becomes an adult but
objects strongly to any addition to the package. With equal validity, we
can turn the argument around and state that additional restrictions may
only imply additional challenges and opportunities for the good
engineer.

A role in this context that private industry must strive to fulfill is to
participate actively in forming opinion when new legislation is being prepared, and to help all the parties involved maintain a realistic perspective. I think that legislation that outruns the present state of technological possibilities is self-defeating. And finally there is one specific aspect
of regulation which should be emphasized by private industry, namely,
standardization, both national and international.

Social Responsibility

The current main issue under this heading is the impact of science and
technology on employment. This issue is widely debated and has called
forth many emotions, as evidenced in the Netherlands, where in 1973 a
coalition government with a socialist majority came into power with an
explicit antitechnology attitude, whereas five years later the socialist-
dominated Workers Council of Philips presented to me the view that
Philips apparently devoted insufficient attention to innovation.

The real danger, however, lies not so much in public discussion as such
but in the possibility that it might lead governments to seek a direct influence in the science and technology policy of individual firms. Two conclusions can be drawn in view of this situation: In the first place, it
should be stressed that not only for industry as a whole is there an urgent
need to explain itself, its mechanisms and its policies, but that also within
that framework it is necessary to explain realistically what benefits can
be derived from further progress in science and technology.

Secondly, it is evident that in most highly developed countries there is an overdose of attention (and financial support) for ailing branches of industry instead of a concentration on the promotion of those branches of industry which carry promise for the future.

Another major social issue in our countries is the mismatch between the outputs of the educational system and the needs of society. I think that there is a real need for serious study of the ways in which society can provide mechanisms to offer a better match between demand and supply in the labor market. This holds not only for society at large but also for the individual enterprise as well. In general, this means that we ought to create further opportunities for qualified work and increased automation to eliminate dirty, dangerous or degrading repetitive work.

A last comment under this heading has to do with certain trends in society brought forward by a small elite, especially in Europe, to enforce egalitarian work situations. I regard such trends as a lethal threat to the continuing progress of science and technology.

Foreign Policy

There are three points I would like to make in the area of foreign policy:

1. The advisability of selling technology to Communist countries is arousing considerable question. Quite apart from the strategic issues involved, one can observe in Europe that in several instances, the vendor of production technology to the Eastern block in latter years has found himself confronted with what he regards as unfair competition. This sort of situation is the especial result of price-setting for exports, which in Eastern countries is determined in an arbitrary way and leads to so-called "political price-setting."

2. The promotion of international standardization remains a prime objective for private industry. In those cases where countries in Europe have gone alone in the establishment of certain standards, the effect generally has been negative. Especially objectionable from the point of view of private industry is the creation of an opposition between technical standardization issues and other foreign policy issues.

3. An important development in the present-day world is the growing resistance to cost-sharing agreements for R&D costs between an international enterprise and its foreign daughters. This had led some governments in emerging countries to disallow the payment of technical assistance fees, a measure designed to promote indigenous R&D. It is not so easy for transnational companies to define their proper behavior under such circumstances, although in countries with a large local market and with a certain tariff protection, product development for local needs is feasible.

Aid to Developing Countries

Aid as such is not a mission for private industry. It is, however, an unde-niable fact, and one that is increasingly recognized in wider circles, that the transnational company is the most effective tool for the transfer of in-dustrial know-how from our countries to less developed countries. At the same time, it should be noted that this know-how is not limited to sci-ence and technology, but comprises many other aspects of business and of infrastructure. Much attention is focused on the issue of transferring science and technology to less-developed countries in view of the forth-coming United Nations Conference on Science and Technology for De-velopment. There is a danger, however, that this conference, instead of contributing in a constructive way, will become a political forum for emotion complaint against the West.

FINAL REMARKS

A number of issues for private enterprise in its science and technology policy have not been touched upon above. I should like briefly to men-tion a few.

(1) The interrelationship between universities as research institutes and private industry remains in most countries vague and undefined. Here is a fertile area for improvement and constructive action.

(2) In view of the willingness of most governments to spend money for the promotion of science and technology within industry, it is regrettable that few effective rules have been formulated for promotion along gener-al lines. The major mechanism today is government funding for explic-itly described projects, with the inherent danger of growing government influence on research programs.

(3) Regulation in general is regarded only as an additional restriction. There are, however, several situations in which new regulation means the creation of a new market, which of course is welcomed by private in-dustry.

(4) Most markets today are well served by established organizations, many of them with powerful, entrenched positions. This implies that market entry demands an excessive price. If one is realistic, this in turn implies that technical diversification can often only be realized either by joining an existing power in the corresponding market segment or by outright acquisition.

I would like to conclude by selecting the two main points from the comments above:

1. In many countries in Europe, there is a tendency within the com-

munity to have government encroach upon private industry with the explicit aim of influencing product policy and its prerequisite science and technology policy. If we realize how extremely different the degree of expertise is within industry and within government, the only advice I can give is to resist this trend as strongly as possible.

2. I want to repeat that I think it is an obligation for industry and for those charged with science and technology policy within industry to explain the underlying economic and technical mechanisms to the outside world.

Workable Mechanisms for Government Action

DUNCAN DAVIES

IMPACT OF TECHNOLOGY ON GOVERNMENT

Introduction

I WOULD LIKE YOU to visualize a sporting event in these parts, two centuries ago in 1779. The great-great-great-great grandfather of one of those present today approaches a bookie and offers a bet. If, he says, there is a United States government-sponsored conference here in 200 years' time, what odds can I have that officers or former officers of George III's great-great-great-great-great niece (who is Queen of England) are freely invited to open and close the conference with their advice? I think he would have been offered a nice bet, like a million to one. And if he had added that the first officer would be a hereditary peer, but descended from someone presently in the money business in Frankfurt, and that the second would explain how England was currently in some trouble through over-concentration on life, liberty, and the pursuit of happiness—and on not imposing taxation without representation—the bookie would have refused to take the money and would have sent for the apothecary and the family to tell them that the bettor was sick. So much for forecasting two hundred years ahead!

Any paper under this title must begin with the observation that the impact of technology on democratic government is far heavier than the impact of government on technology (though this, heaven knows, is important enough). I can paraphrase the title as "the art of the possible" and therefore, by inversion, say that this is equivalent to politics (a subject forbidden to civil servants in the United Kingdom, especially just before or after an election). My only escape is to talk about political theory, which is of course harmless.

The Aim of Democratic Government

The aim of democratic government is (1) to formulate the desires of the

Duncan Davies is Chief Scientist and Engineer, Department of Industry, Great Britain. He was previously Research Director of various Imperial Chemical Industries divisions.

0077-8923/79/0334-0236$01.75/2 © 1979, NYAS

majority as policies and laws; (2) to cause the citizens to conform to these policies and laws by persuasion, rewards, or penalties; (3) to finance the process of government by taxes and borrowing; and (4) to prevent the tendency to corruption conferred by power by periodic reference to the citizens by elections.

The Impact of Technology

Technology affects all four purposes in a democratic state in the following ways:

1. It calls for decisions that are more complex, and therefore makes legislation difficult and less well-understood by the citizen.

2. It gives large numbers of very small groups (for example, computer operators) great power of veto, and therefore makes them less responsive to penalties, which they can collectively resist. It therefore lays greater emphasis on persuasion.

3. It makes the citizen richer and sometimes less resistant to taxation, so that government can become richer.

4. It provides powerful practical arguments for its own exemption from the frequent application of democratic checks and balances, which is evidently inefficient. Alongside similar needs for continuity in policies for education, defense, foreign affairs, health, and so forth, it thus tends to undermine these checks and balances, with obvious consequences.

Technology thus tends to make democratic government (1) more difficult to understand, (2) more difficult to enforce by coercion, and (3) richer.

What Actions Can Government Take?

From this it follows that: (1) the adversary approach is less realistic. (2) If consensus and persuasion fail, there will be an enhanced trend to autocracy. (3) There is a new need for people to have mixed careers, partly as the regulators, and partly as the regulated. (4) The necessity for government in any particular area must be continuously examined and government patterns must become less structured and more flexible.

This last point, of course, is doubly familiar to technologists in business. Let me cite two apposite considerations. First, my teacher was Sir Cyril Hinshelwood, and I clearly remember the essay he had me write on the balance between aggregation and disaggregation: it brought in the change of atmospheric pressure with height, Millikan's determination of all the phenomena of the change of state and the effect of dissolved substances thereon, and heaven knows what else besides. So taking government as crystallinity and entropy reduction and freedom as randomness, gaseousness, and entropy increase, the scientist can see analogies for more dynamic patterns of government. Secondly, businesses centralize when things are tough, in order to conduct existing operations more effectively and cheaply, and decentralize in order to expand freely when there is opportunity. This is how management consultants make money—by pointing out the overshoot in both directions. We have to do the same thing with government.

WHY IS GOVERNMENT BROUGHT IN?

So here is a model that tries to explain why government, as well as being more difficult to conduct, is currently becoming more pervasive. Please note that I am not saying "ought to be;" I am saying "is," and in doing so, I am thinking particularly about France, Germany, Japan, and the United States. The United Kingdom is an interesting case which is going to try to reverse the trend.

FIGURE 1 shows the classical S-shaped curve of the sort that describes the development of a successful product, or business, or industry, or national economy. It plots the consumption, or output per head of population against time. (The curve can be converted to a straight line by plotting $\log [f/(1-f)]$ against time, where f is the fraction of the process completed at time t). The figure is divided into three parts. The first, marked

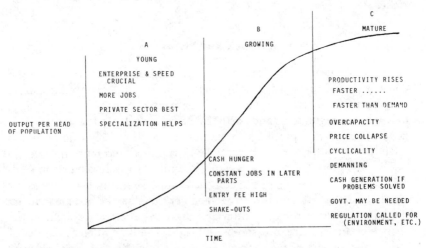

FIGURE 1. Development of an industry.

A, describes the phase of development in which activity is young and accelerating. There is a need for uninhibited division of labor to solve problems with the minimum of bureaucracy and interaction. Profitability is
high, and profits are growing because a social need has recently been
met. There is enough cash either internally generated or readily
available, for cash hunger is not yet very large. The second, marked B, illustrates the state when activity is growing fast. Competition is severe, so
that margins are less; cash demands are high, and borrowing may be
more difficult. The pressure for competitors to retain market share to
lower costs may create overcapacity and generate price collapse and the
shake-out of the weaker competitors. The entry fee for new entrants is
becoming prohibitively high. And the number of jobs, which was growing in the "young" phase A, is probably stabilizing. Thus, in the mature
phase C, the survivors recover from the traumas of phase B, and productivity then tends to rise faster than per capita demand. If external export
markets can expand, employment can be maintained; as soon as this
levels off, the industry or business may need to reduce employment in
order to maintain efficiency.

There is, of course, a classical answer to the problems of maturity: to
generate new industries or businesses and thus absorb the skills made
available from a mature industry within a new activity. Such a policy
was pursued for many years by the electrical and chemical industries.

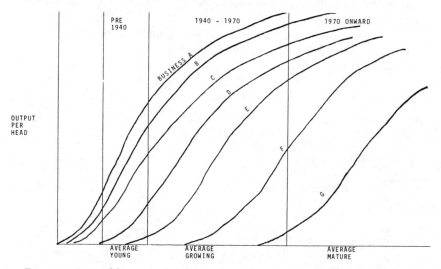

FIGURE 2. A possible reason for the increase of government activity in some industries.

The latter, in fact, was introducing scientifically-based new products and systematic manufacture into industries whose technology was craft-based—soap, fertilizers, materials, textiles, pharmaceuticals, and agriculture. However, here too, after a spectacularly close-spaced series of S-shaped curves, it becomes harder to find new worlds to conquer. This is shown schematically in FIGURE 2. At the heady start, youth is at the helm, and methods of the uninhibited private sector are at their very best. As time goes on, the "average" business becomes a cash hungry, growing affair, and the industry becomes very much concerned with difficulties of finance. Then the average becomes "mature" (although of course there is still the occasional "young" business).

Government climate-creation seems to be far more important in these later stages than in the earlier. Standards and quality control matter more and regional employment problems tend to arise. Hence, without any ideological change at all, mixed-economy methods are brought in. But it is important to observe that selectivity, however achieved, is far more important in the smaller economy than in the top nation, which can still manage to have competitively adequate market shares in a wide range of businesses and industries, and thus avoid the need for selection. A particular difficulty for the small economy today is that industries and businesses interact; a prosperous automobile industry is assisted by prosperous machine tools, components, robotics, electronics, and materials industries. Readily available effort near at hand is better than imports from another economy.

Hence the growing role of government and the growing concern about its bureaucracy and intervention.

TABLE 1

WHAT ACTIONS CAN GOVERNMENT TAKE?

1. Specify and purchase { products / technology
2. Generate* { products / technology
3. Operate { manufacture / distribution
4. Finance*
5. Legislate, regulate, standardize, plan
6. Promote (patent system, etc.) or inhibit (secrecy laws, taxes) technology transfer
7. Educate and train

* The problem areas

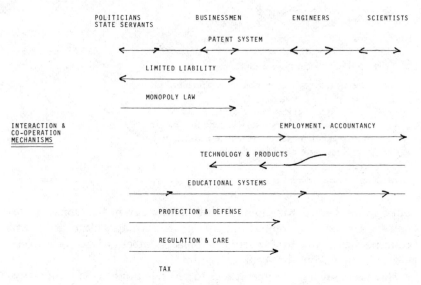

FIGURE 3. Interaction and cooperation mechanisms.

What Actions Can Government Take?

TABLE 1 seeks to list the possibilities for action and FIGURE 3 illustrates the way in which these possibilities span the activities of different participants in the business/technology/science/government system.

The most significant point in present experience is that the interaction has been becoming more detailed. The initial devices for promoting innovation (the patent system), commerce (limited liability), and for preventing excessive aggregation of power (monopoly law) were very general in character. Technology was seen to be assisted by publicity and inhibited by secrecy; it was therefore rational to offer a temporary, 16-year monopoly to an inventor provided that his work was original and not obvious, and that he published his working procedure. This required no government action other than an examination to ensure that the rules had been observed, and the operation of a legal system to permit challenges. Government sponsorhsip of education and training could and did include measures to help technology: the United States land-grant colleges, the German and Dutch Technical High Schools, and certain of the French *Grandes Ecoles* exemplify this.

In a somewhat more specific way, but one which still finds acceptance, governments can stimulate technology by their purchasing policies. Many commercially important developments have so arisen: gas tur-

TABLE 2

WHY IS GOVERNMENT BROUGHT IN?

1. National Security or aspiration
2. Needs that the free market finds difficult to satisfy:
   (a) "patient" money
   (b) general services as of right (e.g., education)
   (c) employment or prestige beyong the "natural"
   (d) regulation against immediate market interest
   (e) concentration of effort → monopoly
3. Measures requiring coercion

bines in the United Kingdom, nuclear technology in the United States and the United Kingdom, and highly productive agriculture in several countries have come from wartime purchases. Integrated microelectronic circuitry came from the Apollo program.

However, when governments seek to generate products or technology, or to operate technology, manufacturing, distribution, or finance instruments, or to legislate constraints, serious difficulties arise, and these must now be examined.

WHY IS GOVERNMENT BROUGHT IN? AND WHERE DO PROBLEMS ARISE?

The answer to these two questions are indicated in TABLES 2 and 3. It is clear why government involves itself in national security aspirations. There are also areas, such as nuclear power or long-term resource development, where money requirements involve a delay between investment and reward that the free market finds difficult to stomach. Sometimes there is a conflict between the interests of the citizen as purchaser and as neighborhood resident; there is therefore a need for regula-

TABLE 3

WHERE DO PROBLEMS ARISE?

1. Absence of clear index of efficiency
2. Absence of comparative standards and competitive pressure
3. Absence of political continuity, necessitated by democratic checks and balances
4. Bureaucracy from { probity
   { concentration and coordination
5. Inertia and inflexibility (applies to other large organizations)

TABLE 4

WHAT IS AGREED TO WORK?

1. *Technology transfer* stimulation by patents
2. Invention (stimulation by patents)
3. *Enterprise* stimulation by limited liability
4. *Rapid development* by wartime programs
5. *Pursuit* of any *clear consensus* (e.g., the Apollo project)
6. *Quality improvement* through standards
7. *Health, safety, and welfare* in areas of consensus, by regulation and standards
8. *Some* areas of *education* (e.g., land-grant colleges)

tion to support neighborhood needs in order to bring them into balance with consumer strength. Similarly, if there is to be coercion or penalty, they can only be applied broadly by government; one aspect of such coercion may be action against monopolies to avoid the stifling of competition.

One basic requirement for government action is a high degree of popular consensus such as exists in wartime, or lends general support to education and the suppression of proven toxicologic hazard. When it is absent, as when the government is being selective and helping one industry or region and not another, there must obviously be dispute. And another is a reasonable measure of efficiency. As the implementer of consensus laws, government may be able to steer clear of the detail that tends toward bureacratic confusion and strangulation. But where there is detail, there tends to be difficulty. Hence, the areas where government action works smoothly, or on the other hand, is controversial, is summarized in TABLES 4 and 5.

One may ask whether government action in technology is massive enough to worry about. TABLE 6 shows clearly that it is, and that the extent is similar in several countries, although the pattern differs.

TABLE 5

WHAT IS CONTROVERSIAL?

1. Selective policies based on long-term views
2. Legislation and regulation in disputed areas
3. Generation of products and technology by government for market
4. Manufacture and distribution by government
5. Work based on "science push"

NOTE: Selection is needed everywhere, except for the top nation, for which the need is much less.

TABLE 6

How Much is Done Now by Government in Science and Technology?

	Gross Expense of R&D		Where Executed (%):			How Paid for (%):		
	$1000M	GDP (%)	Busi-ness	Gov-ern-ment	Edu-cation and other	Busi-ness	Gov-ern-ment	Miscel-laneous
United States (1975)	34.6	2.3	68	16	16	43	54	3
Germany (1975)	8.85	2.1	66	16	18	53	45	2
Japan (1975–76)	8.77	1.7	64	13	23	65	16	19
France (1975)	6.00	1.8	61	23	16	40	41	19
United Kingdom (1975–76)	4.65	2.1	63	27	10	41	52	7

TABLE 7

Conclusions

1. There are many areas of science and technology in which government is agreed to have workable mechanisms.
2. Government agencies can execute quickly and forcefuliy where there is consensus and will.
3. Difficulties lie in the areas of choice, control, and resource economy.
4. Where choice is controversial, it helps to separate it from execution (customer-con-tractor), but good dialogue must be maintained.
5. Controversial choice is helped by close partnership with the private sector.
6. Improvement in control and resolve economy is helped by private sector involvement.
7. Government is needed alongside mature industries when productivity grows faster than demand.

Conclusions

We can sum up the situation as indicated in TABLE 7 as follows:

How a balance between private and public action is achieved in a democracy, is a matter for voters to decide. There is evident inefficiency in the centrally controlled states, and there are now hardly any cases of industrial economies operating under a condition of complete market freedom. The trends are hard to predict, but when the decision has been made concerning the extent and pattern of government action, there is a great need for sensitive and intelligent collaboration between government and industry on the uses and application of technology.

A Summary Perspective

WILLIAM D. CAREY

To REVIEW THE ISSUES presented at this Conference, it is useful to keep in mind the context that has strongly influenced the quality of our discussions, a context affected by the troubling performance of the United States national economy, declining productivity, a high regulatory profile, a favorable government climate for scientific research, a social passion for accountability, and an undisputable centrality of government's role in shaping the outlook for science, for better or worse.

Let me touch generally upon the flavor of the discussions that have been held during this Conference. We felt that the base strength of the science enterprise is strong and that it is not burnt out, even though our postwar structural arrangements, keyed to a government-*cum*-university partnership, are struggling and may call for a new look. We thought that the critical paths of applied science and innovation are clouded by regulatory overload. We sensed that the management of our political economy does not adequately value the potential of research and development, and that there is a disjunction between economic policy and science policy.

We concluded that technological problems are now central rather than peripheral in the context of economic development, and that classical economics no longer will be adequate to solve future problems of resource allocation. We seemed to agree that public opinion and the focusing of ideas of social responsibility are changing the ways by which science policies are worked out, and that this means that decisions are harder to come by as social interventions multiply. The transition to service-type economies puts more stress on science to improve productivity and to improve benefits relative to cost. With all this, the dilemmas of explaining science policy to the general society are becoming more troubling. The assumption that science and technology operate in a free-flowing market seems to be inappropriate. Increasing research may turn out to be of lower priority than getting ourselves straight on incentives, disincentives, and barriers.

William D. Carey is Executive Director of the American Association for the Advancement of Science. He was formerly Assistant Director of the Bureau of Budget, Executive Office of the President.

0077-8923/79/0334-0245$01.75/2 © 1979, NYAS

There is a great sense of confusion over the dual roles the Federal government plays as chief patron of science and at the same time the chief hairshirt of innovation. We also sensed that public overexpectations for science-based solutions could prove embarrassing in the coming years. On the other hand, some of us take comfort in the belief that pluralism is alive and well in this country and is winning out over centralism; this cheerful opinion came through in the panel session on social responsibility in science. Enthusiasm was somewhat dimmed by the view that the terms of science's social contract are undergoing renegotiation: science now extends beyond its cognitive role to a role of publicly accountability as questions of value are thrust into debate.

Although the Conference was billed as international, most of our time was spent in arguing the American situation. International issues were not ignored, however. They emerged in the sessions on foreign policy and development assistance, and the discussion touched on problems of deterrence and defense along with strategies to deal with the North-South relationship. The term "aid" is inadequate, because it carries with it the dream of new markets, and we preferred the word "cooperation." We thought that the use of classifications like "developed" and "developing" was not always helpful.

The opinion emerged that the real problem of the emerging nations is the lack of infrastructure: technically educated people, organization, and related factors. And "technology transfer" received some criticism when, in practice, we fail on our side to see that the transferred know-how must be enriched by some new knowledge created in the receiving country. On the whole, the discussions on foreign science-policy problems persuaded us that there is need for much more policy research capacity, not just for the enigmatic question of development but equally in the hope of getting at measures of developmental progress and good analysis of strategic weapons systems policy.

The area of telecommunications was debated in depth and with sophistication. There was strong feeling that government regulatory practices are at cross purposes with the public interest, and that deregulation would increase competition and trigger innovation. But competition will necessitate abandoning some of the ideas of equity on which current pricing practices are based, thus generating new political dilemmas. The panel found itself disagreeing over the net additional social value of fast-approaching expansion of communications capacity, but there was consensus that more education is needed on what the coming technological advances will be. It was clear, in any case, that the economics and politics of communication will be as significant in the next five years as will be technological innovation.

The panel on the economy dwelt on signals of adverse trends relative to productivity and innovation, noting that the decline in productivity happens to go along with decreasing R&D spending in the United States. Lack of venture capital for risk purposes was cited, while the group noted some dissatisfaction with the present set of innovation indicators (in terms of adequate reflection of the benefits of innovation). The point was made that policy changes designed to spur investment in R&D alone would not suffice, and that what we need are incentives to stimualte long-term investment in all the essential elements in the flow-process of innovation.

The panel on materials policy called for more basic research and greater support for demonstrations of materials processing. It was thought that more knowledge of the interplay among energy, environment, and employment is needed to shape policies for materials discovery and processing.

The issue of environmental policy permeated all of the discussions, which is hardly a surprise since science and technology policy would not be as complex without taking into consideration questions of value and measurement. We emphasized the function of science in strengthening the quality and confidence of the regulatory regimes to which we are committed on behalf of environmental quality, and by inference we expressed less than great confidence in the existing regulatory quality.

"Regulation" was a term heard repeatedly during this Conference and the adjectives used to describe it were mostly uncomplimentary, such as "inconsistent," "drawn-out," and "inflexible." This indicates a troublesome gap separating public administration from science and research. But hope was offered for a new trilateral and nonadversary institutional approach that would bring industry together with government and academia to look for better ways of managing regulation. There was a strong argument made for involving scientists and technologists in all levels of the regulatory system, especially in the legislative sector.

As for the problems of food, the panel felt general satisfaction with current levels of research and advanced technology, but saw a disastrous absence of an overall food policy to provide direction. An adequate basis must be found for establishing a relationship between nutrition and health status, identifying the relations between consumer patterns and health, better education of the public, and a public policy that supports a free market in defining and meeting food needs.

The energy predicament received due attention. The panel decided that science and technology policy for energy should be formulated on a systems basis, in which a choice is made among alternatives with full awareness of comparative risks and social benefits. A fundamental precondition of this process is the use of science policy to get better information on the social and political implications of different policy choices. This involves better ways to get the public to participate and to provide for cross-communication among groups at interest. At this point in the discussion the argument for international coopoeration in energy policy was strongest.

To return to the question of social responsibility: the panel felt that the relationship between science and the American society is in the process of being renegotiated. With this is the concern to scientists to accept that in the future there will have to be involved, at least partly, in political controversy. However, the point is unclear at which a line is drawn between the responsibility of the individual scientist and that of corporate science.

If we concluded this colloquium with the impression that science policy in the United States is long on questions and short on answers, we would be justified. But it is asking too much to expect straightforward answers where science and public policy come together. Such straight-line answers are seldom encountered in *any* of the critical zones of national policy, whether the subject is economic stabilization policy, growth policy, health-care policy, defense or arms-control policy, monetary affairs, or energy policy. So why should science and technology policy be any different. in fact, it is likely that if we *had* a straight-line science policy, we wold be in a worse situation than we are in now.

From my own perspective, there is much to be said for the American preference for the empirical approach to science and technology policy, as distinct from a generic or inclusive approach. If we glance at some other fields where the framework does have generic aspects, we see that the outcome is not necessarily beautiful or good. There is, for example, welfare policy, a tortured and unsatisfactory system that operates under national rules. Then there is tax policy, necessarily an inclusive class of policy, about which not too much needs to be said by me. We could consider, with eyes averted, the area of transportation policy, as to which the interstate area of transportation policy, as to which the interstate commerce clause of the Constitution has given rise to a tangle of law and regulation from which we may never extricate ourselves. And there is food and drug policy, where generic concern for the public interest has produced over time a unidirectional bent that is perhaps the prize model of straight-line decision-making unhampered by counsels of discretion.

So it occurs to us that our impatience with sidewise and fits-and-starts

policy-making for science may be ill-considered. Perhaps we should set-
tle for a consensus climate for science, a climate in which some rain will
fall and in which not all the sessions will be in the sun. I submit that a
"rational';' policy for science is very likely to be one-sided, deriving from
government and public policy. It could be something else if the structure
of United States science reflected normal checks and balances, permitting
the negotiation of policies . It should be quite clear by now that we do
not have such a structure, and there is no likelihood of getting one. Last
week, in the Letters section of *Science*, a correspondent could say "In the
process of becoming the principal source of research funding, the federal
government has exercised a subtle but pervasive control over the kinds of
research that are performed. As one member of a funding agency
remarked to me. 'We admire innovation, but we don't trust it. And we
fund what we trust.'" There we have it.

At this symposium we have ranged across the expanding subject of
policy problems. We see that if science policy remains essentially reactive, there will be no chance to use or lead-time to accomplish what we
must through science and technology. What will be left are "crash"
responses to overtake situations that have gotten out of hand. This is
substantially where we are at in the energy situation, and it is where we
are likely to be with climate, the management of ocean resources, and
materials. It would be greatly to our advantage to focus science and
technology policy on providing alternatives to the critical resource supply
shortages that the United States economy is bound to run up against in
the coming two or three decades. And such a policy would supply the
strategic element that is missing from our empirical approach to science
policy as contrasted with the present *ad hoc* attempts at solving problems.

Although we seem, at this conference, to be fairly busy on an agenda
of science policy issues, we might reflect upon Russell Peterson's exercise
last year in "vacuuming up" so to speak, a whole universe of issues as a
base for setting priorities for the Congressional Office of Technology
Assessment (OTA). As reported recently in *Nature,* the OTA asked more
than 5,000 persons to name the critical technological issues facing
America and the world. That query produced 1,530 issues, and a diligent
search of the literature found 2,875 more, for a total inventory of 4,405
potential OTA study topics. The OTA settled on 30, leaving only 4,375
to be attended to elsewhere. What strikes us, in the face of such numbers,
is the incommensurability between the scale of the issue "universe," on
the one hand, and the prevailing level-of-effort in science policy
research. I suggest that as long as the gap remains so great, we cannot ex-

pect much except improvisation, muddling, and misfiring along the route to science policy. If a fraction of the attention that we give to the administration and accountability of science could be transferred to policy research, we might find our bearings sooner.

As in the case with just about every variety of issue, paradox is with us. On the one hand, each of us here approaches the issues of science policy with a mind that is not as open as we'd like it to seem. (I apologize if there are any saints in the gathering.) So when the United States delegation goes to Vienna in August 1979 for the United Nations Conference on Science and Technology for Development, we have to understand that it is not a band of apostles bearing the new truth but an instructed delegation that brings a point of view that in time will contest with a decidedly opposite point of view. The same holds for meetings on the Law of the Sea, nuclear nonproliferation, SALT, climate, population, and the rest of the "smoking gun" issues in which science policy must come to terms with national self-interest. It may not be pretty, but it is reality. And it so happens that there is another large difficulty—the time constant because the making of science and technology policy is not independent of the tide of events and interventions, which take the matter out of our hands. So if science policy is to make a difference, it must not strive for elegance (which runs the risk of pretentiousness); rather, it should seek to be decision-forcing, and it must work with one eye on the clock.

Science policy in the third postwar decade is a rather different matter than it was in the first and second decades. We had things generally our own way then, and the notion of limits had scarcely entered our mind. There was a strong mandate for science, and we saw the future as an endless horizon. There was a brief but great age of building. Confidence ran high, and magnanimity was its child. It was the time when the government saw its role as the supportive agent of basic and applied science and education for science. The policy agendas took note of nuclear science, medical research, geophysics, strategic warfare, and then space. They gave little heed to energy supply, environmental hazards, the biosphere, nuclear proliferation, and regulation of research; further, the sacrament of accountability had not yet been discovered, there was no Group of 77, and the conscience of science was not yet aroused on behalf of human rights.

We have come far and fast to the realities of the *third* postward decade of science and science policy, to the economics of constraint, the politics of tension, the hyperadministration of research, the dilemmas of social and ethical responsibility. Now, at last we appreciate that technology is power and has to be watched closely. If so much could change from the first postwar decade to the third, what is ahead for the fifth and the sixth decades? We may not consider straight-line policy making for science to be what the doctor ordered, but we may end up getting it if we have nothing to put in its place. This is the large task that faces us as we deepen the responsibilities of policy centers like the one sponsoring this Conference.

Science and Technology
Problems and Opportunities
at the State Level

HUGH L. CAREY

J. ROBERT OPPENHEIMER wrote in 1953, "The open society, the unrestricted access to knowledge, the unplanned and uninhibited association of man for its furtherance—these are what makes a vast, complex, ever growing, ever changing, ever more specialized and expert technology world, nevertheless, a world of human community."

Through the development scienceand technology we have combatted disease, lit our streets and our homes, and improved the productivity of our farms and the efficiency of our industries. We have developed new forms of energy and materials, improved local and worldwide communications, learned to harness the resources of our environment, and most recently, have reached beyond the moon towards the farthest expanses of our solar system.

But, as Henry David Thoreau wrote in 1848, "In the long run, men hit only what they aim at." Our aim in the past two decades has been low.

According to a 1977 report of the United States Department of Commerce, technological innovation was responsible for 45 percent of the nation's economic growth from 1929–1969. High-technology industries have had productivity rates twice as high as those of low-technology industries. Real growth rates have been three times as great. The high-technology industries have had nine times the employment growth and only one-sixth of the annual price increases of low-technology industries. Yet while the crucial role high-technology development plays in our economy is demonstrably clear, it is also depressingly evident that its growth has been drastically reduced.

In 1968, from 300 to 400 high-technology industries were founded in the United States. In 1976, none were founded. We endured a period that

---

Hugh Carey is Governor of New York State.

0077-8923/79/0334-0253$01.75/2 © 1979, NYAS

has been characterized by the breakdown of United States innovation. This is undeniably a dangerous trend for any nation made great by the fortuitous confluence of physical and human resources. It is a waste of the innovative abilities of the citizens upon whom our nations have depended for growth, prosperity, and progress. Innovation is the engine that creates jobs; it boosts productivity and contributes to a strong and healthy balance of trade. Innovation is the engine that drives many companies to the front of their industries. A slowdown in innovation inevitably leads to a slackening in the nation's economic growth, which an industrial country cannot afford. This slowdown can and must be avoided by a return to our innovative roots.

But we cannot sit back and wait for this to happen. "Discovery is not invention," said Thomas Edison, who believed that inventions were not random strokes of luck, rather he thought that they were the products of purpose. Discoveries were accidents, and great innovators have never been able to wait for such accidents to occur. We have to go in search of, and organize for, innovation as we would have to to reach any other goal. Edison pursued his beliefs with the formation of the Edison Light Company in 1878, thus incorporating the "invention factory" into the American economy.

And today the greatest hope for increasing the rate of productivity also lies in advances in technological innovation. This can be achieved only through organized research and development. We require a renewed technological innovation on a scale not achieved in two decades. Not only is this renewed activity crucial to our future economic growth, but it is also fundamentally related to our health and survival. Science and technology policy must meet the twin challenges of growth through innovation and environmental quality.

Industrial growth and technological development in the delicate ecologic balance of the modern world has created vast opportunities. But it has also spawned new problems, such as those of energy supply, air quality, water quality and supply, elimination of toxic wastes, and materials shortages. We must recognize that continued progress relies upon scientific and technological advances to restore the ecological balance.

Government must take a lead in revitalizing the innovative activities of the past. In the twentieth century, the industrial research laboratory has played a prominent role in innovation. Such diverse inventions as the jet engine, the gyrocompass, the helicopter, magnetic tape recording, the zipper, and the self-winding wristwatch have emerged from independent inventors. Government must play a major role in creating the proper economic climate for the "invention factories" of industrial research.

And government must work to encourage the continued participation of the enterprising individual whose meager resource may be inadequate to perfect his device in the face of establishment indifference.

Here in New York State, long a national and world leader in the development and application of science and technology, we feel a keen sense of obligation to meeting those challenges. We have taken new steps to honor this obligation. Here we have an unmatched reservoir of human and physical resources. We have nationally recognized research universities and institutes, both public and private. We have a strong existing base of research and development industries ranging from the computer industry, electronics and communications, through medical science and chemistry, to aerospace and metallurgy. We are a pioneering state in energy development and pollution control. Within our boundaries are dynamic young companies, firms that are international leaders in research and development. Our universities educate more than 10 percent of all American doctoral scientists and engineers. And we have one of the most highly skilled labor forces in the world. But most of all we have a talented and committed citizenry, ready and able to join with government to seek solutions to the problems confronting our state and nation.

As Albert Einstein said in 1931, "It is not enough that you understand about applied science in order that you may increase man's blessings. Concern for man himself and his fate must always form the chief interest of technical endeavors." The way to ensure that concern for man is foremost in our minds is to inspire the involvement of our citizenry.

We have begun this drive by revitalizing the New York State Science and Technology Foundation to provide a central focus for our efforts to stimulate the development of high-technology industries. This foundation seeks to encourage an improved environment for the advancement of high-technology industries in New York State. Another prime focus is to identify and aid inventors and entrepreneurs in the commercialization of new products and processes. We have created an Energy Research and Development Authority (ERDA) to help solve our energy problems. Its primary role is to accelerate the eventual commercial application of technologies required to meet our energy needs.

Industry cannot function in a competitive world without reasonably priced and environmentally safe energy. Yet New York State, like the nation, is excessively dependent on foreign energy supplies. The solution to these problems is important to the State's existing economy, and the further conduct of energy research also requires the employment of qualified people. This produces jobs and ultimately a new industrial system to manufacture and market what is developed. In this regard, ERDA has developed a 455-acre energy research park at the Saratoga Research and Development Center. More of these parks are planned. Thus, by turning to science and technology to solve our energy problems, we are also promoting economic expansion.

New York State has also begun a renewed effort to encourage scientific and technological development to solve other problems and to lay the groundwork for future expansion. Last summer, I directed my Economic Affairs Cabinet to form a task force on high-technology opportunities. This task force is mandated to coordinate State agency activities to stimulate the growth of high-technology industries. It must develop policy initiatives to promote increased activity in research and development. I have also called upon the talents of our private sector and formed an Advisory Council on high technology to work with the task force. Composed of men and women of demonstrated abilities and leadership in the business, scientific, and academic sectors, the Council has been working with the task force to devise strategies to solve complex problems of mutual concern.

We in New York are embarking again on a journey to the endless frontier of science. And as Thoreau so aptly wrote, "The frontiers are not east or west, or north or south, but wherever a man fronts a fact."

"This world is but a canvas to our imaginations, dreams are the touchstones of our character." That statement was believed to be made by Arthur D. Little, the pioneer of the private consulting laboratory. In 1921, he set out to prove that in modern America anything was possible. Inspired by the folk saying, "You can't make a silk purse out of a sow's ear," he obtained ten pounds of gelatin made from those appendages. He spun an artificial silk tread from the gelatin, wove that thread into a fabric, and actually produced an elegant purse. If American science and technology can make a "silk purse out of a sow's ear," it can also confront the challenge of renewed innovation to tackle the problems of today's society.

Through science and technology we can meet the needs of our people. But government cannot play a passive role. It must engage in policies designed to aid and spur innovation to open the gates of innovative ingenuity. The "breakdown of United States innovation" must not be permitted to describe the 1980s. This generation of Americans has inherited a tradition of resourcefulness. As we face new obstacles and challenges, we must call upon that tradition. We must develop the policies to ensure that science and technology continue to be a part of the future, and not a memory of the past.

Role of Congress in Science and Technology Policy

DON FUQUA

I WILL DISCUSS here the many dimensions of science and technology policy and their impact on America from the perspective of our national legislature, the United States Congress.

The scientific progress of modern man has been the most significant factor in alleviating the burdens of primary existence to make life easier, healthier, and happier. And yet we are witnessing the growth of a dedicated antiscience, antitechnology movement in today's society, at a time when the institutions that have served us so well—our universities, our industry, and the Federal government—are facing increasing problems.

It is somewhat ironic that as some in the United States entertain an antitechnology mood most of the rest of the world—including the developing countries—is embracing science and technology as a key to their future development.

As a nonscientist who has been involved with science, science policy, research and its applications as technology, I have had the advantage of two perspectives. On the one hand, I have been the beneficiary of scientific innovation, using its many wondrous products in my daily life, pausing only occasionally to consider how these things have pervaded my existence.

On the other hand, I have had the perspective of a Member of the Committee on Science and Technology, where I have worked closely with scientists and engineers to formulate policy and legislation that either mandates or regulates scientific research and its technology outgrowth for the benefit of the nation and its people.

It is from this second perspective that I know we must always have the freedom to pursue knowledge. Inherent in this freedom, however, is the responsibility to use such knowledge constructively. The exercise of this responsibility is as important as our right and freedom to inquire, to gain new knowledge, and to create something new from it.

Yet there is a potential conflict which may result from the collision of

Don Fuqua is a member of the United States House of Representatives and is Chairman of the Committee on Science and Technology.

0077-8923/79/0334-0257$01.75/2 © 1979, NYAS

differing philosophies. Scientific freedom is not necessarily accepted in various parts of the world. Moreover, in areas such as oceanographic research, various nations have placed major constraints on the conduct of research within their 200-mile economic zone. These constraints stem from the perception that various Western nations have benefitted to the detriment of developing countries by conducting research in their waters. In some developing countries the publication of such research results is opposed because it might reveal important natural resources such as oil.

Science and technology are neither good nor bad; rather the manner in which man uses them determines whether they are good or bad. We need not discard technology because we have failed to use it perfectly, but we must assume more of the moral responsibility that accompanies the use and accessibility of this technology. Herein lies the guts of science policy and the direction in which modern society will move. And herein lies many of the solutions to the world's most basic problems, such as food supply, adequate energy resources, and good health.

Our emphasis upon moral responsibility concerning technology leads again to another kind of collision with nations of the developing world. For example, seeking solutions to some of the problems I mentioned raises questions about how what we have learned about environmental pollution and industrial safety should be applied in the developing world. The issue is a difficult one. On the one hand, the United States labor movement by and large opposes the movement of United States firms abroad, where they operate under less stringent regulations and standards than exist in this country. On the other hand, many nations in the developing world place a much lower priority on environmental control than on industrial and agricultural development. There are some who suspect any attempts of ours to encourage stronger environmental controls as simply another form of trying to keep the lesser developed countries down.

Science policy encompasses basic research, applied research, the funding and regulation of each, and technology development, demonstration and assessment. These all fall within the jurisdiction of the Committee on Science and Technology. Although establishing and directing science policy is in itself a formidable task, this policy is just one part of a dynamic network of issues that must be coordinated and integrated to meet our national needs and the requirements of our foreign policy.

In broad generalization, America seems to have evolved into a nation specializing in leading-edge technology. This is for all practical purposes our "technological frontier." In fact, the presence or conception of a frontier has shaped our history since the beginning. For roughly the first hundred years, we had the geographical frontier of uncharted, unsettled land toward which we stretched our most daring dreams and adventures. With the close of this frontier, the energy and imagination that took us 3,000 miles across the land broke out in a new direction.

The second hundred years of our history were marked by unprecedented scientific technological and industrial expansion. Since World War II, the pace of this expansion has actually catapulated us beyond the planet Earth. All frontiers are typified by high risk and this "technological frontier" that we have pursued is not only accompanied by high risk, but also by high costs for expensive equipment implicit in necessary trials and errors.

Less developed countries often do not understand the complexity and rich variety that has characterized economic and technological progress in the United States. It all looks so easy in retrospect—and because we have succeeded so well there has until very recently been little examination and study of our great engine of progress. There is, I believe, one very major understanding that we must attempt to convey to the developing world: science and technology in themselves are not the sole explanation for the kind of progress they see in the United States, Japan, and Western Europe. This progress results from perhaps indefinable mixtures of political structure, societal forces, attitudes about work, the type of education system, and the like. The point is that there are many influences at work and sorting them all out is a very difficult task.

Even in our country we are beginning to understand that we must reexamine our own system and the influences at work. For example, as we start this third century of our common history, we have come to recognize that however innovative leading-edge technologies are at the onset, they mature and the technological know-how becomes common knowledge. At this point, other nations have available, without risk, the information that enables them to produce a finished product that can successfully compete with our own. We have seen this occur most recently with such things as color televisions, citizen band radios, and tape recorders. Some experts suggest that this is becoming a pattern for America and the problem has increasingly attracted the attention of industry, academia, and government.

Attempting to understand the complexities of this issue and to discover viable solutions is difficult because this situation represents a new experience for America. When a nation is an acknowledged leader in technology or anything else, there is little reason to explore what it is that has allowed that country to maintain its position. Indeed, its position is usually taken for granted. However, when that prevailing posture is challenged or undermined, it becomes very important to analyze those things contributing to the new trend. We have only recently come to understand that this is a task we must tackle.

We are now discovering contradictions between the intent and the actual results of many of our own policies. To cite just a few:

• For inventions developed with government funds, the innovation incentive of patent protection is often undermined by our policy of placing these patents in the public domain.

• Government-industry cooperation in large research and development (R&D) projects of national concern is encouraged at the same time that Federal patent policy sometimes discourages this cooperation.

• We call for technological innovation while tax and regulatory barriers slow down that innovation.

• We wish to foster cooperative industrial R&D on high-risk, expensive projects to alleviate national problems, but our antitrust laws discourage this cooperation.

• Technical assistance to less developed countries raises their competitive capability in international markets such as textiles, soy beans, and other areas where the United States once had an unquestioned lead.

I am, however, encouraged by recent developments that I will mention specifically. In 1976, the National Science Policy Organization and Priorities Act became law. This bill, which originated in the Science and Technology Committee, is landmark legislation in the sense that it puts into law the government's intent to eliminate "needless barriers to scientific and technological innovation."

But the role of the Congress is by no means limited to the drafting and debate of legislation. We have an equally significant role to play in overseeing the administrative implementation of these statues. We intend to carry out this responsibility with respect to the Science Policy Act, giving particular attention to Title III. Among other things, this title specifies the need for a regular review of Federal regulations and practices that may inhibit inventiveness in the private sector.

President Carter has expressed repeated concern over the state of research, industrial innovation, and our technological posture in the world. Shortly after his inauguration, he appointed Dr. Frank Press to head the Office of Science and Technology Policy. As a further step in this direction, the Administration's recent budget message to the Congress requested a 5 percent real increase in research spending, while most budget categories experienced considerable austerity.

A White House Domestic Policy Review Panel is now completing six months of intensive work under the direction of the Department of Commerce in examining the spectrum of United States technology with special emphasis on technological innovation. The recommendations to the President are sure to address some of the policy contradictions I just mentioned and will be followed up by appropriate Executive and legislative action.

As part of this recent activity and inquiry, the Science and Technology Committee will shortly publish a compendium of papers entitled, "National Science and Technology Policy Issues." It is a compilation of views by individuals in industry, government, and universities concering the relationship between technology and the economy. This document lays the groundwork from which many of our future activities are likely to take shape.

During the first week in April 1979, the Science Committee will hold a series of hearings to review the total Federal R&D budget, which for fiscal 1980 amounts to more than $32 billion. By the definition of the Office of Management and Budget (OMB) itself, this represents between 15 and 24 percent of the "relatively controllable" portion of the entire Federal budget. We expect to have a broad range of witnesses, including R&D experts from industry and professional societies, the President's Science Advisor, appropriate directors of OMB, and the Comptroller General.

Science and research have traditionally flourished in America's colleges and universities. Academia has always been the seedbed for research activity, but there was a dramatic increase in this activity beginning with the Second World War. From pre-World War II to the present, we have also seen an equally dramatic change in university funding, which has made these institutions increasingly dependent upon the Federal government for support. Back in 1930, about one-half of the funding for American universities came from endowments. These play a much smaller role today, averaging from 8–10 percent of their total funding.

At present, for state universities, the major source of funding is the state government, and for all universities, the Federal government is a heavy participant through project grants. However, state funding formulas for university support are closely tied to enrollments and these enrollments are on the decline. In the late 1950s and 1960s, when university research funding grew rapidly, the primary rationale for this increased support was major growth in enrollment. It is going to be difficult for institutions to request an increase in funding from their state legislatures, including funding for research and science teaching, in the face of diminishing enrollments.

At the same time, Federal funding for all universities has leveled off because inflation and general budgetary paring. This in turn has placed restrictions on the ability of all our universities to hire the bright, young Ph.D. scientists who so often come up with the new theories and ideas on which science is nourished. We are in danger of experiencing a "missing" generation of scientists. This is going to place the universities, the research community, and the state and Federal governments at a juncture where decisions regarding both the future financing of our universities and support of the nation's basic research programs will have to be reexamined.

Let me now turn to specific features of science and technology in international development. Any discussion of science and technology in today's world would be incomplete without a prominent mention of its international dimensions. I recall several years ago when Dr. Edward David of the Exxon Corporation appeared before our Committee as a witness on the proposed Science and Technology Policy Act. He said that "one of the three major problems of science policy in the years ahead would be the relationship of science and technology and international relations." Most certainly this is what has happened in the world of the late 1970s as we approach the 1980s.

As just one small sample of my own and my Committee's involvement in recent months, I could note the following:

• In 1978 the Committee on Science and Technology worked cooperatively with Chairman Zablocki of the House Foreign Affairs Committee on enacting Title V of the Foreign Relations Authorization Act of fiscal year 1979. Called "Science, Technology, and American Diplomacy," this title gives the State Department a strengthened and important role in scientific and technological matters. Then-Chairman Teague and I supported this legislation as being complementary to the Science and Technology Policy Act of 1976.

• In February 1979, Chairman Zablocki and I cochaired a Congressional seminar on the United Nations Conference on Science and Technology for Development. (We are both Congressional advisors to the United States delegation to the Conference.)

• In early spring of 1979, Congressman George Brown and I testified before the Foreign Affairs Committee of the House in support of the newly proposed Institute for Scientific and Technological Cooperation. If the Institute is established, it will likely become an important facilitator for focusing more sharply scientific and technological capabilities on development processes in various nations.

Without underestimating the difficulty of the problems I have discussed, I might conclude on a somewhat optimistic note. Science, by its very nature, has always transcended territorial borders, and scientists, along with other scholars and artists, have perpetuated the idea of the universality of mankind through their work. The day is also long departed when technology can be confined within national boundaries. Widespread education, instant communication, and the knowledge that man is dependent upon the bounty of his planet for survival, have drawn us all into an ever-tighter circle. In our urgent search for new sources of energy, in our exploration of space, and in our growing understanding that all nations, large and small, developed and developing, need the fruits of science and technology to survive, we have come to cooperate and share as well as compete with each other. The worldwide challenge of the new scientific and technological frontier is common to us all and we must face it together. I do not believe that national goals of competitiveness in the marketplace contradict our international goals of an improved standard of living for all peoples. I do not believe that technology has ever weakened our spirit, but rather that it has eased our toil and helped in our unending search for new knowledge.

A Labor Perspective on
Science and Technology

WILLIAM W. WINPISINGER

THORSTEIN VEBLEN PREDICTED that the rise of scientists and engineers into corporate management positions would lead to conflicts between the manager's quest for profits and the scientist's devotion to technical efficiency and conservation (doing more with less). Veblen felt that these conflicts would have beneficial results that would transform the social power structure.

In this he appears to have been wrong. Instead, for the most part, scientists and engineers have either wittingly or unwittingly, willingly or unwillingly, placed their expertise at the service of the corporate state, a productive system efficient in details, but supremely wasteful and irrational in its general tendencies.

This failure is not an indictment of scientists, but an observation that the notion of technological determinism (technology shaping society) is a myth. Far from revolutionizing society, technological changes merely reinforce the existing distribution of power and privilege.

Max Weber was closer to the mark when he observed that modern technology at the service of the corporate state would weave a paralyzing web of instrumentality that would enslave modern man. For the corporate state has come to assume the guise of modern technology, the management experts lending to the power of capital the sanction of science.

Not only the actual machinery of production, but also the entire bureaucratic operation of corporate enterprise has taken on the look of an efficient well-oiled mechanism—the very embodiment of technical reason—against which opposition could not but appear irrational.

There are three important further observations to be made in this regard.

1. Scientific knowledge is not ever apolitical (i.e., neutral) in its application. For example, when new machines are designed through the use

William W. Winpisinger is International President of the International Association of Machinists and Aerospace Workers.

0077-8923/79/0334-0264$01.75/2 © 1979, NYAS

of a particular new technology, there are design paths that can be followed that would strengthen the power of the workers, and there are design paths that weaken the power of the workers and their unions. Since technology is in the hands of the corporation, you can imagine which design paths are followed.

2. Scientific knowledge at the service of the corporate state assumes the purposes of the corporate state.

3. When competition is eliminated through concentration of economic power, the role of innovation through the application of scientific knowledge diminishes in significance as a tool for maintaining corporate power, just as do all other costly or risky devices for enhancing efficiency. The resulting slowdown of technological advancement is just another manifestation of the general stagnation that results from economic concentration.

In this manner, we come to the issue of the independence of the corporate state versus the assertion of public policy considerations for the common good, as it relates to innovation. However, this seems hardly to be the issue that fits the facts. The issue is rather, how to reassert the public rights as preeminent over the ambitions of the corporate state while it may still be possible.

Innovation has an enormous impact on our lifestyle and standards. Edward Denison of the Brookings Institute has estimated that "advances in knowledge" were the biggest single sources of national economic growth from 1929 to 1969. Data Resources, Inc., a respected economic consulting firm, has found that companies that invest heavily in research and development increase employee productivity 75 percent faster than all manufacturers. They also create jobs 120 percent faster, while raising prices only one-fifth as fast.

But industries' technical resources are being moved away from long-term basic research, toward short-term projects to improve existing products. The amount of total basic research performed by industry dropped to just 16 percent in 1976 from 38 percent twenty years earlier (15 percent in 1977). This decline has allowed other nations to chip away at America's technological leadership. It has been estimated that as much as two-thirds of worldwide research and development is now conducted in foreign laboratories. Technology has been a leading American export for several decades, yet this nation's trade surplus in R&D-intensive goods has been eroding steadily for several years.

A philosophy that growth through acquisition is preferable to growth through innovation seems to prevail. Large corporations capable of generating cash tend to hold it or to use it to acquire other firms. Small companies with innovative ideas find it difficult to locate investors willing to risk their capital to bring their ideas to market.

More than half of current United States research and development spending is for defense. Moreover, much of the remainder is being devoted to areas related to compliance with legislative and regulatory requirements. Since real overall industry research and development spending has not increased, this means that proportionately less is going into truly innovative research and development.

By contrast, during the past ten years, Japan increased its research and development spending by 74 percent. Japan now has more scientists and engineers working on nondefense research and development than the United States has.

In 1977, federal government funds were the source of two-thirds of all money spent for basic research, and government funds amounted to approximately half of that spent at the applied research and development stages. Approximately 65 percent of total government spending for research and development is spent through contracts to industry. But in the United States, more and more of us are beginning to recognize that science and technology research and its development to practical applications is another one of the very crucial areas of our society and economy that must be reexamined and reoriented in light of its contribution to maintaining and furthering monopolistic corporate power.

By controlling the means of production, the corporate state already wields enormous power. As a result of increasing numbers of mergers and acquisitions beginning after World War II, this power has been concentrated in fewer and fewer hands. The number of merger transactions involving $100 million or more nearly doubled in 1978, and total dollar volume rose to $34.2 billion.

A major result has been the tightening of corporate control over markets and prices through the elimination of effective competition. This is the basic reason why attempts to control inflation through policies aimed at suppression of demand are ineffective in today's world. Such policies no longer put effective pricing pressure into the market because effective competition has been systematically eliminated. Any potentially effective competition that arises is bought up or driven out of business by temporary market manipulations.

At the end of World War II, the 200 biggest manufacturing firms had 45 percent of all assets of United States industry. Today it is 60 percent. In 1960, some 450 firms controlled about 50 percent of the nation's total manufacturing assets. Today it is 70 percent. Less than 1 percent of American manufacturers control 88 percent of the industrial assets and receive more than 90 percent of the net profits of industrial firms in the American economy. Just as short a time ago as 1960, small and medium-sized businesses controlled 50 percent of the nation's corporate assets. That figure had dropped to 30 percent by 1976. With such concentration of economic power, does anyone wonder how big business continues to get away with escalating prices and profits regardless of demand?

The resulting economic control wielded by the corporate state is consistent with its major purpose of maximizing returns on investment while minimizing risk.

Although at times we have lamented business's lack of consideration for the public interest, most Americans have generally agreed that as long as society had available adequate means of asserting the public interest when required, its application of economic control otherwise yielded desirable efficiencies. However, now the waste and inefficiency produced by economic concentration and the lack of genuine competition are creating serious roadblocks to our attempt to deal with new economic problems associated with the growing scarcity of resources.

Genuine competition occurs only when sellers are too numerous to cooperate to fix prices. Unhappily, this condition no longer exists for a major portion of the American economy. This has the same unsatisfactory result on competition for innovation that it does on price competition. This is why the majority of significant product and technology advancements have achieved abroad in recent years.

The major question we face in our society today is what to do about the fact that the concentration of corporate power has grown to the point where it is now immune to the traditional attempts of governments to reassert authority. It is even knowledgeably questioned whether this concentration of power, when coupled with the apparent corruption and indecisiveness of government, is not already so extensive and pervasive that the balance of power has shifted to the point where the conglomerate corporations cannot be brought under control.

This view is somewhat understandable when we realize that the self-insured glut of corporate profits provides unlimited resources for lavish political contributions, extraordinarily lobby expenses, political and business bribery and kickbacks on an international scale, progaganda campaigns, front group foundations and institutional advertising—all to advance narrow corporate interests.

Every American is affected by the discretionary decisions left to corporate officers: what to produce, what price to charge, where to locate.

Of major importance among many decisions left in corporate hands are decisions about research and development—how much to spend, the appropriate allocation separately to research and development, the appropriate rate of introduction of new products or improvements. When the threat of competition has been removed, the result is a shifting of emphasis with respect to research and development allocations. This shift is away from pure research and development of new products into enhancement types of development of already existing products, and even into marketing research or phychological inducement through advertising.

Our monopolistic multinational firms receive a majority of the government funds for research and development. Between 60 and 70 percent of research in the United States is performed in laboratories of private companies. So the benefits of government investment in research should transmit directly to industry. Even most federal laboratories are operated by major companies under service contracts. Thus, they have been able to see to it that they get the bulk of government funds allocated for research and development. And there are strong indications that these firms merely use government funds for research and development to displace corporate funds rather than to increase the total research and development effort. Industrial research today is dominated by a small number of very large corporations.

The top 10 percent of those firms in 1976 performed almost 70 percent of the total United States research and development effort. Ten firms accounted for more than 36 percent of all expenditures that year. Small firms get less than 4 percent of government outlays for research and development.

Yet a recent unpublished study by the Office of Management and Budget (OMB) credited firms having less than 1,000 employees with almost half of the industrial innovations between 1953 and 1973. These are the firms the conglomerates are gobbling up to get the new technologies they need, but cannot get, spending all the government's research and development money.

These firms, which make up the corporate state, having dissipated the publicly financed United States technological lead by transferring it all over the globe for a profit have noticed that the technical lead is going fast. But they have developed the true welfare mentality.

They encouraged the Carter Administration to set up a Domestic Policy Review on Industrial Innovation to be conducted by some twenty-eight federal agencies, with various advisory committees, the major one being a Business Advisory Committee headed by William Agee (Secretary Blumenthal's successor at the Bendix Corporation) and otherwise made up of representatives from the business roundtable. This com-

mittee submitted its report in January, after having met for a total of only seven hours to discuss the issues.

The report consists of the most blatant expression yet of the corporate
state's "hidden agenda." Not only do the committee members want the
government to increase public investment in research and development,
but they further want it to reduce long-term capital gains taxation, increase tax investment incentives and rollback health and safety and
pollution regulations—all in the name of making research and development a better investment for corporations.

While a number of the recommendations made by the Committee are
supposed to benefit small businesses, they will actually be of greater
benefit to giant corporations, as was vehemently pointed out by the one
small-business representative on the Committee.

It has been reported to me that a high Commerce Department official
associated with the Domestic Policy Review, when asked if the government intended to continue to reserve a portion of its research and development dollars for small business, replied affirmatively, stating that
acquisition of these firms seemed to be the only way of getting new technology into the larger firms. That certainly shows where this Administration's priorities lie.

No one has a measure of the amount of scientific knowledge that is being withheld from development by the corporate state, but it is knowledgeably reported to be substantial. Market statistics suggest that
technology is way ahead of American industry's willingness to absorb it.
And I am further convinced that such a withholding is consistent with
their prime objective.

We have today, on every hand, evidence of industry's resistance to
alternative technologies that might endanger their control.
Technologically oriented companies are bought up not only to exploit
new technology, but often to suppress it or to delay its fruition.

Many small businesses have technology sitting on the shelf, so to
speak, but are limited in venture capital to bring its fruits suitably to
market. Yet they cannot interest the larger firms or the banks. Foreign
corporations show more interest.

Our major corporations claim that the risks involved do not make
development a good investment. Even if this is so, what about all of the
proven technologies being used by our international competitors that
have not been applied by United States firms? These inventions could go
a long way toward closing the technology gap.

We see further evidence in the automobile industry's reluctance to
develop smaller cars that are more fuel-efficient, or to more vigorously

develop alternatives to the oil-driven piston engine, or to meet pollution standards that at least one foreign competitor has met with ease through the development of the stratified charge engine.

Furthermore, we see the general discrediting and delaying tactics surrounding the development of solar energy technologies. The oil companies, which already own huge coal reserves and exercise virtual control over the uranium industry, have recently been obtaining a substantial interest in the copper industry and in solar technology companies, while at the same time conducting a major campaign to discredit the sun as a viable alternative energy source.

One would like to think that their motivation was simply not to be left out should their assessment prove erroneous. However, the result is that they gain the means of controlling the development and the rate of introduction of these technologies.

With such control, these companies can sit on this technology until they have exhausted the potential of their investments in more conventional energy sources and technologies.

You need not be convinced that this is happening to see that something is very wrong, that it is possible, and that under existing law, there is little that we can do to stop it.

However, it is interesting that of all industry groups whose expenditures for research and development are reported, fuel-providing companies spent the least—only 8.2 percent of profits. Mobil Oil reported no expenditures for research and development in their 1977 10K Report to the Securities and Exchange Commission.* †

---

* In the 1977 SEC 10K Report filed by Mobile Corporation, the expenditures for research were not identified separately or itemized. However, some of Mobil's research activities are described on pages 1–16. The 1978 report states: "research expenses for the year were $86,000,000 in 1978 and $73,000,000 in 1977."
—EDITOR.

† [Note added in proof: This simply emphasizes again what is well known; that a corporation's cost accounting techniques (particularly, we have learned, in the oil industry) are a matter of internal convenience, depending upon whether they are "composing" their "presentation" for themselves, the stockholders, Internal Revenue Service, the Securities and Exchange Commission or the "people." This assures the necessary flexibility for all *post hoc, ergo propter hoc* rationalizations. Even using the newly minted figures in the 1978 10K, Mobil's 1977 R&D was only 7.3% of profits, and we don't know but that it was all devoted to Mobil's nonfuel subsidiaries such as Montgomery Ward or Container Corporation of America. Everybody knows Mobil has PR Plenty—but they aren't piping it in to heat my house yet.—AUTHOR]

Let me give another example of this corporate obstructionism: I believe that answers to some of our problems of protecting and enhancing life on this planet lie in the area of developing the potentials of space exploration and use. These potentials range from environmental protection through eventual removal of polluting industrial processes from the earth's atmosphere, space manufacturing, and totally new manufacturing processes and products that could be developed in zero gravity, to unlimited cheap power through an array of solar power satellites.

Most of the basic research and a good deal of the design work has already been accomplished. Yet it has not been possible to convince the National Aeronautic and Space Administration (NASA) or Congress to proceed at anything other than a financially starved, minimum-feasibility study, snail's-pace rate with respect to exploiting these potentials.

I think we can see the hand of corporate America in this decision, from the power companies and the oil industry, to the companies that make up the traditional military-industrial complex and the major conventional aerospace industry. They all feel threatened by the potential transformation that could result. These are powerful opponents of scientific and technological developments.

Innovation means disruption and inconvenience. As companies become larger, richer, and more insulated from the necessity to adjust prices to demand fluctuations, they start to trade off innovation for convenience and efficiency for stability. Apparently, without the spur of competition, industry, with a few notable exceptions, will not continue to invest in new technology.

Recent years have brought increasing recognition that we live in a finite world with finite resources. In one sense this is good, because it represents the truth, even if we misperceive where the boundaries are. Furthermore, it encourages a justified concern and respect for the appropriate use of all of our resources. But such knowledge can be bad when it is used to frighten us into acceptance of limits of freedom and potentials of development and to foster the notion that solutions should not be sought because there are none.

I view scientists, researchers, and designers as the true conservationists. The whole history of the advancement of science and technology can be capsulized as "learning how to do more with less." I do not believe that the limits of this advancement have been reached. But I do believe that the corporate state has managed to tighten its grip on research and development results so that they now, more than ever, serve to advance the interests of the corporations. And these interests are not compatible with what is needed to strengthen our economy and our society. So we are encouraged to believe that the advancement of science and technology has faltered, or that we have placed too much faith in science's potential.

Increasing size alone translates into increased market power; market power at some point translates into the capability to administer prices or negotiate for capital. That power is roughly equivalent to the government's power to tax in its potential to destroy and disrupt. If government does not put a stop to this trend, we will soon be governed by the largest corporations.

The corporate state grown to its present power has many economic tools with which it maintains its position. Perhaps the most significant of those tools is the withholding of investment for growth. Market power is concentrated in finance as well as in manufacturing. In 1977, our five largest banks held 42 percent of the total deposits of our fifty largest banks.

We are at present in the midst of a capital strike. With their monopolistic positions, corporate and bank investment decisions are not made just on the basis of maintaining market position. Without effective competition that position is already secure. Therefore, investment decisions can be made or not made on the basis of further consolidation of economic and political power and weakening potential rivals, or simply keeping workers, governments and the population at large in a dependent state of insecurity and uncertainty.

Therefore, we find corporations withholding investment in new and more efficient plants and equipment. With no competition and an assured market and control over prices, the enhancement that greater productivity would bring is less critical to sustaining the enterprise's economic position. Withholding investment capital puts pressure on the economy and therefore on the government to obtain favorable regulatory, legislative, and financial allocation decisions.

The result of the total situation is that corporate America will not take investment risks either on new plants and equipment, the development of new products, or on the furtherance of scientific research. This is because they do not need to do so because they have concluded that the return on such investments is not high enough with respect to the risks involved.

And, further, they have learned that there are power leverage advantages with respect to the allocation of their investments. Therefore, investment today is largely limited to high-return short-term investments. Thus, we have a capital strike as serious economically as a national labor strike would be. What it amounts to is an escalation of class warfare, perpetuated not by revolutionaries of the masses, but by an elite managerial class of corporate boards of directors.

If we have, in fact, entered a new period in the history of life on this planet where scarcity of resources is a prominent feature, then we must become more prudent in the use of all of our resources—not just the natural resources, but also the resource of knowledge that results from scientific research. Decisions about the support of scientific research and particularly the development of possible technologies and products can no longer be left strictly in the hands of a few directors of conglomerate America, any more than we can permit them to keep a complete lock on production, prices, plant location, and investment.

We often think of corporations as being sacrosanct, quais-autonomous entities, and we allow ourselves and them to forget that they exist basically as creations of the state. If in their actions they fail to make a positive contribution to the common good of the state, their charters to do business could be pulled by the state.

Corporations have an obligation to do more than simply maximize the return to shareholders. The state continually redefines, through all kinds of laws and regulations, the exact nature of how the corporation must operate to serve the common good. Somewhere along the way, we have allowed the notion to creep in that there are definite limits to how far the state can go without infringing on some god-given rights of the corporation. But, there really are no limits on the requirements the state can place on a corporation to remind it of its obligation to serve the interests of the state's citizens or to indicate where those obligations may override stockholder interests. And I, for one, believe it is time to do a little more reminding.

Unless we move quickly to establish stronger mechanisms for public control over investment decisions and credit allocation, power will continue to become even more concentrated in the hands of a few—in short, corporate priorities may overwhelm all human considerations. In a period of stagnation and slow growth, the cost of natural resources is going up, and we can expect that the value that is placed on human life is going to go down. Already we see a new antidemocratic, antihuman ideology emerging under a banner of fiscal soundness.

What is most needed for the sake of restoring jobs, economic health, increased productivity and an improved climate for research and development is a restoration of competition in the marketplace before it is too late. To help accomplish this, we need legislation such as that proposed by Senator Kennedy to ban mergers between large corporations. Also several worthy bills have been introduced calling for divestiture by major oil companies of their holdings in other sources of energy. Other bills are needed to confine corporations' activities to their proper economic role, and to insure that they meet their social responsibilities.

In January a Presidential Study Commission (The National Commission for the Review of Antitrust Laws and Procedures) reached similar conclusions when it issued a unanimously endorsed report containing more than 50 recommendations to expedite complex antitrust cases and to increase competition in the economy. The recommendations include the following: (1) amending the Sherman Act to make it easier to prove companies are attempting to monopolize markets; (2) urging Congress to determine whether the government should be able to sue to break up large corporations having "persistent monopoly power" without having to prove deliberate anticompetitive conduct; and (3) enacting legislation requiring regulatory agencies to give greater weight to competitive considerations in their decision-making.

I feel that a part of the obligation a corporation owes to the public is to produce the best and cheapest product possible with the most advanced technology it has available. Upon failure to develop a promising new product or technology, the corporation should lose its exclusive rights to any patents involved. A corporation should not have the power to delay or withhold the fruits of research and development. A procedure whereby petitions by interested parties for the setting aside of patent restrictions or forced licensing should be instituted in instances where compelling evidence indicates that the patent owner has not made a reasonable attempt to develop all promising applications.

It is, of course, important to keep adequate funds flowing for research and development. Scientific discovery is not a spigot you can turn on or off at will and get results. The support should be there all the time. But the public does require some practical resuls, especially since so much of the funding is out of the public treasury.

The only way we can really get this rolling again is to recreate an environment where the thrust of competition forces corporations to take the risks inherent in developing new products and technologies.

This will require a broader-scale effort than simply encouraging government and corporations to work more closely in the area of research and development. It involves a struggle against the stagnation-producing economic security of corporate America. Understandably, they will fight back, and they have a great deal of power.

As you see, I hardly view the issue of asserting the public interest in research and development policy as a question of preserving the mutual independence of the government's legislative authority and financial resources and of industry's role as a developer, investor, and marketer, because I believe that industry already exercises far too much independence or dominance in the decision-making in this area, and that government has far too little.

One does not consider the appropriate application of the Marquis of Queensbury rules when the big guy is beating the hell out of the little shrimp. Afterwards is time enough to establish the rules and appropriate handicaps for future altercations.

The issue is not free market versus big government and mutual independence. The government is already deeply involved in our research and development decisions and will inevitably become more so. The issue is really the larger matter of what kind of government, what kind of involvement, and most importantly, on whose behalf.

The answer is already clear if we allow the corporate state to continue to have its way.

APPENDIX

Lewis M. Branscomb
Vice-President and Chief Scientist
International Business Machines Corporation
Armonk, New York 10504

Arthur M. Bueche
Senior Vice-President, Corporate Technology
General Electric Company
Fairfield, Connecticut 06431

Hugh L. Carey
Governor of New York
Executive Chamber, State of New York
Albany, New York 12224

William D. Carey
Executive Director
American Association for the Advancement of Science
Washington, D. C. 20036

Umberto Colombo
President, Comitato Nazionale per l'Energia Nucleare
00198 Rome, Italy

Lee L. Davenport
Vice-President and Chief Scientist
General Telephone and Electronics Corporation
Stamford, Connecticut 06904

Edward E. David, Jr.
President, Exxon Research and Engineering Company
Florham Park, New Jersey 07932

Duncan Davies
Chief Scientist and Engineer
Department of Industry
Abell House
John Islip Street
London SW1P 4LN, England

Bernard Delapalme
Directeur de la Recherche Scientifique et Technique
Elf Aquitaine
75739 Paris CEDEX 15, France

William Dill
Dean, Graduate School of Business Administration
New York University
New York, New York 10006

* Program participants often served as panel members for more than one panel.

Merril Eisenbud
Professor and Director,
Laboratory for Environmental Studies
Institute of Environmental Medicine
New York University Medical Center
Tuxedo, New York 10987

Herbert I. Fusfeld
Director, Center for Science and Technology Policy
New York University
New York, New York 10003

Don Fuqua
Member, United States House of Representatives
Rayburn House Office Building
Room 2268
Washington, D. C. 20515

J. E. Goldman
Senior Vice-President and Chief Scientist
Xerox Corporation
Stamford, Connecticut 06904

Philip Handler
President, National Academy of Sciences
Washington, D. C. 20418

Norman Hackerman
President, Rice University
Houston, Texas 77001
Chairman, National Science Board

N. Bruce Hannay
Vice-President, Research and Patents
Bell Telephone Laboratories
Murray Hill, New Jersey 07974

James L. Hayes
President and Chief Executive Officer
American Management Associations

Walter R. Hibbard, Jr.
Department of Engineering
Virginia Polytechnic Institute and State University
Blacksburg, Virginia 24061

D. Bruce Merrifield
Vice-President of Technology
The Continental Group
New York, New York 10017

Dorothy Nelkin
Professor, Program on Science, Technology and Society
and Department of Sociology
Cornell University
Ithaca, New York 14850

Richard R. Nelson
Department of Economics
Yale University
New Haven, Connecticut   06520

Rodney W. Nichols
Executive Vice-President
The Rockefeller University
New York, New York   10021

Grace Ostenso
Technical Consultant
Subcommittee on Science and Research and Technology
Committee on Science and Technology
United States House of Representatives
Suite 2321, Rayburn House Office Building,
Washington, D. C.   20515

A. E. Pannenborg
Vice-President
Board of Management
N. V. Philips Company
Eindhoven, the Netherlands

Harvey Picker
Dean, Faculty of International Affairs
Columbia University
New York, New York   10027

Gerard Piel
President and Publisher
Scientific American
New York, New York   10017

Frank Press
President's Science and Technology Adviser
Director, Office of Science and Technology Policy
Executive Office of the President
Washington, D. C.   20500

Sherman K. Reed
Vice-President and Director for Chemical Technology
FMC Corporation
Philadelphia, Pennsylvania   19103

Lord Rothschild
N. M. Rothschild & Sons Ltd.
London EC4P 4DU, England

John C. Sawhill
President, New York University
New York, New York   10012

Elmer B. Staats
Comptroller General of the United States
Washington, D.C. 20548

Klaus-Heinrich Standke
Director, Office for Science and Technology
United Nations
New York, New York 10017

David Swan
Vice-President, Environmental Issues
Kennecott Copper Corporation
New York, New York 10017

Norman Terrell
Deputy Assistant Secretary for Science and Technology
Department of State
Washington, D.C. 20520

Monte C. Throdahl
Senior Vice-President
Monsanto Company
St. Louis, Missouri 63166

William W. Winpisinger
International President
International Association of Machinists and Aerospace Workers
Washington, D. C. 20036

PANEL PARTICIPANTS

J. C. Agarwal
Director of Development, Ledgemont Laboratories
Kennecott Copper Corporation (became Vice-President, Engineering, Charles
  River Associates Inc. as of April 9, 1979)

Dominique Akl
Scientific Coordinator
United Nations Conference on Science and Technology for Development

Laurence Berlowitz
Assistant Vice-President for Academic Affairs
New York University

Reynald Bonmati
Vice-President, Research and Development
Elf Petroleum Corporation, U.S.A.

Sidney Borowitz
Executive Director
New York Academy of Sciences

Robert L. Brass
Director, Business Systems Analysis
Xerox Corporation

Richard H. Braunlich
Special Assistant to the Vice-President for Chemical Technology
FMC Corporation

Alfred E. Brown
Director, Scientific Affairs
Celanese Corporation

David N. Campbell
Vice-President
Computer Task Group, Inc.

Jean Cantacuzène
Le Conseiller Scientifique
French Embassy
Washington, D.C.

Donald W. Collier
Vice-President, Technology
Borg-Warner Corporation

R. W. Diehl
General Manager, Technical Marketing
The Continental Group

Ted Dintersmith
Department of Engineering-Economic Systems
Stanford University

Dennis Dugan
Chief Economist
General Accounting Office
United States Government

John M. Dutton
Professor, Graduate School of Business Administration
New York University

Martin C. J. Elton
Chairman, Interactive Telecommunications Program
School of the Arts
New York University

Felix J. Germino
Vice President of Foods Research
Quaker Oats Company

Arthur Gerstenfeld
Professor and Head, Department of Management
Worcester Polytechnic Institute

T. Keith Glennan
Former Member, Atomic Energy Commission
Former Administrator, National Aeronautics and Space Administration

Ralph Gomory
Vice-President and Director of Research
International Business Machines Corporation

Kenneth Gordon
Head of Planning
National Bureau of Standards

Robert I. Hanfling
Executive Assistant to the Deputy Secretary
Department of Energy

Bernhard Hartmann
Professor
Technische Universität, Berlin

Richard F. Hill
Executive Manager
The Engineering Societies Commission on Energy, Inc.

Milton B. Hollander
Vice-President, Technology
Gulf + Western Industries, Inc.

John Holmfeld
Specialist in Science and Technology
United States House of Representatives

Paul F. Hopper
Corporate Director, Scientific Affairs
General Foods Corporation

David A. Irwin
Chief of Policy and Rules
Federal Communications Commission

Lionel Johns
Assistant Director
Office of Technology Assessment
United States Congress

Donald D. King
President
Philips Laboratories Division
North American Philips Corporation

Ursula Kruse-Vaucienne
Program Officer
Division of International Programs
Directorate for Scientific, Technological, and International Affairs
National Science Foundation

Gösta Lagermalm
Senior Policy Adviser
Swedish Board for Technical Development

Richard Langlois
Department of Engineering-Economic Systems
Stanford University

Charles F. Larson
Executive Director
Industrial Research Institute

Susan Leech
Assistant to the Vice-President
Carnegie Capital Advisors
Prescott Ball and Turben

John Lightstone
Union Carbide Corporation
*   and Assistant Adjunct Professor of Finance*
Graduate School of Business Administration
New York University

John Logsdon
Director, Graduate Program in Science, Technology, and Public Policy
The George Washington University

Franklin A. Long
Henry Luce Professor of Science and Society
Cornell University

Steven Marcus
Technology Review
Massachusetts Institute of Technology

Thomas McCarthy
Director, Research and Development Department
The Procter and Gamble Company

Mary Mogee
Analyst, Science Policy Research Division
Congressional Research Service

S. L. Meisel
Vice-President, Research
Mobil Research and Development Corporation

Mitchell Moss
Associate Professor, Planning and Public Administration
Graduate School of Public Administration
New York University

Henry R. Nau
Associate Professor of Political Science and International Affairs
The George Washington University

Arthur Norberg
Division of Policy Research and Analysis
National Science Foundation
  and Head, History of Science and Technology Program
University of California, Berkeley

Richard B. Opsahl
Director, Technical Liaison
Grumman Aerospace Corporation

J. E. Penick
Senior Vice-President
Mobil Oil Corporation

Andrew H. Pettifor
Policy Analyst
Division of Policy Research and Analysis
Directorate for Scientific, Technological, and International Affairs
National Science Foundation

Rolf P. Piekarz
Group Leader
Division of Policy Research and Analysis
Directorate for Scientific, Technological, and International Affairs
National Science Foundation

Geoffrey Place
Vice-President, Research and Development
The Procter and Gamble Company

Herman Pollack
Research Professor of International Affairs
The George Washington University

S. Victor Radcliffe
Senior Fellow
Resources for the Future

Umberto Ratti
Scientific Counselor
Italian Embassy

R. R. Ronkin
Staff Associate, Division of International Programs
Directorate for Scientific, Technological, and International Affairs
National Science Foundation

John M. Rozett
Office of the Governor
State of New York

Allen S. Russell
Vice-President, Science and Technology
Aluminum Company of America

Lewis H. Sarett
Senior Vice-President for Science and Technology
Merck and Company, Inc.

A. George Schillinger
Dean of Management
Polytechnic Institute of New York

Vivien B. Shelanski
Staff Director
Science and Society Program
New York Institute for the Humanities

Benjamin S. P. Shen
Reese Flower Professor and Chairman,
Science Policy Committee
University of Pennsylvania

Allen M. Shinn
Senior Science Associate
Office of the Director
National Science Foundation

Paul Silverman
President, Research Foundation
State University of New York

Bruce Smith
Director, Policy Assessment Staff
Bureau of Oceans and International Environmental and
*   Scientific Affairs*
Department of State

Ralph H. Smuckler
Director
Institute for Scientific and Technological Cooperation

Lowell W. Steele
Manager, Research and Development Planning
General Electric Corporation

Robert Stibolt
Department of Engineering-Economic Systems
Stanford University

Martin Summerfield
Astor Professor of Applied Sciences
Director of Graduate Studies
Graduate School of Arts and Sciences
New York University

John Thompson
Counsellor, Science and Technology
British Embassy

John F. Tormey
Director, Corporate Technology Policy
Rockwell International

Frank N. Trager
Professor of International Affairs and Director, National Security Program
New York University

Charles Weiss
Office of Science and Technology
The World Bank

Jean Wilkowski
United States Coordinator for the United Nations Conference
*   on Science and Technology for Development*
Department of State

Max L. Williams
Dean, School of Engineering
University of Pittsburgh

FUSFELD + HAKLISH, eds   Science +
technology policy.

SCIEN___D

This bo

14 OCT 19